Man
and
Meteorites

MAN AND METEORITES

Brian Pejovic

Editors
T. H. Stewart and S. M. Stewart

Thames Head

First published in Great Britain 1982
by Thames Head Limited
Avening Tetbury Gloucestershire

Copyright © Thames Head Limited 1982
Text copyright © Brian Pejovic 1982

ISBN 0 907733 01 8

Designed by
Playne Design Avening Gloucestershire

Typeset in Century on a Compugraphic Editwriter by
Maurice McKee Cirencester Gloucestershire

Printed in Great Britain by
R J Acford Chichester Sussex

Contents

Acknowledgements

My grateful thanks go to the authors of the many scientific papers, monographs and studies which I consulted during the preparation of this book. Museums that afforded their generous co-operation are too numerous to name, but I must particularly thank the National Research Council of Canada – for data and photographs about the Canadian Meteorite Observation and Recovery Project – and the Centre for Meteorite Studies in Tempe, Arizona. My thanks also to all those who so kindly lent me their photographs.

Dr H. H. Nininger – a legend in his own lifetime – has shared with me his reminiscences covering sixty years of meteorite research and recovery.

Stan and Louise White and Miss Ulla Fitz have given notable advice and encouragement, and dedicated librarians in every continent where I have gone meteorite-prospecting have been prodigal in their assistance.

I thank them all

B. Pejovic
Geneva – June 1981

Stars falling from the sky

A contemporary artist's impression of the crucially significant meteorite shower falling on L'Aigle, France in 1803. The accounts of unimpeachable eye-witnesses caused the powerful French Academy of Science publicly to accept the fact that bodies of extraterrestrial origin could bombard Earth.

Foreword

In Riga, in 1794, a paper published by E. Hladni proposed to a derisive scientific world the cosmic origin of meteorites. Ten years later Thomas Jefferson – scientific scholar as well as President of the United States – publicly denounced the claim by two reputable Yale professors that "stones have fallen from the sky."[1] "I would rather believe," said Mr Jefferson "that those two Yankee professors would lie than that stones could fall from Heaven."[2]

But the professors did not lie, and centuries of observation confirming the fact were not to be denied for much longer. On April 26, 1803, a shower of meteorites fell in L'Aigle, France, over an area of several square kilometres. An investigation conducted by the French Academician and astronomer, Biot, and the reports of irreproachable witnesses were so overwhelming that the powerful French Academy of Science had to accept the fact that bodies of extraterrestial origin *could* fall from the sky. Thus, the last barriers of prejudice gave way, and the science of meteoritics was born.

Now, nearly two hundred years later, we are irrevocably committed to interplanetary travel, and the search is on for other life in the Universe. The minute study of meteorites is now known to provide vital clues to this extraterrestrial treasure-hunt. These fragments coming from the immensities of space offer decoding possibilities to tantalise astronomer, mathematician and geologist alike. And even to those of us not trained to interpret their messages, they offer more besides – the unimaginable beauty of their celestial fireworks, and the sheer excitement of, just possibly, finding our own fallen star.

We can now examine the stars and planets through super-powerful telescopes. The space shuttle will launch a satellite whose own telescope will improve our viewing a hundred-fold; and by the end of this century a lunar base may be providing us with an entirely new perspective on the Universe. But meteorites, 'out of this world' in very truth, are in our hands now; and many more wait to be found and identified, with all their clues still unread as to the cosmic events which led to the formation of our planet and gave it life.

If my book can convey even a little of the fascination of this science so newly claimed from the field of fantasy, and can share with you my own delight in it, I shall be well content.

[1] The highly respected Judge Wheeler wrote an eye-witness account of these 'falling stones' after he saw a fireball riding over the horizon on December 14, 1807, at 6.30 in the morning.

[2] The meteorites of this fall of 1807 are now on display at Yale University.

To the memory of my brother

Some of the beauty and the untold secrets of the universe are captured in a meteorite

Venerated since prehistoric times, 'shooting stars' still stir the imagination of man. They are objects from the ocean of space that can be held in the hand and examined – tangible evidence of the universe beyond us, and a continuing source of enlightenment as to its mysteries.

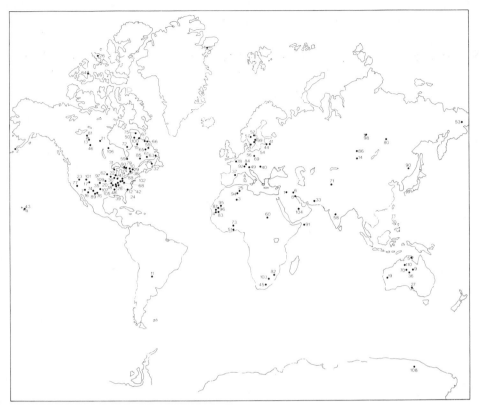

LOCATIONS OF TERRESTRIAL IMPACT STRUCTURES

The structures indicated include confirmed, probable, and suspected meteorite impact craters.

1 Al Umchaimin Crater, Iraq
2 Amak Island Crater, Alas.
3 Amguid Crater, Algeria
4 Aouelloul Crater, Mauritania
5 Arnhem Land Crater, Australia
6 Baghdad Craters, Iraq
7 Barringer Crater, Ariz.
8 Basra Crater, Iraq
9 Boxhole Crater, Australia
10 Brent Crater, Canada
11 Campo del Cielo Craters, Argentina
12 Carolina Bays, U.S.A.   .
13 Carswell Lake structure, Canada
14 Chinge site, U.S.S.R.
15 Clearwater Lakes, Canada
16 Crater Elegante, Mexico
17 Crestone Crater, Colo.
18 Crooked Creek structure, Missouri
19 Dalgaranga Crater, Australia
20 Decaturville disturbance, Missouri
21 Deep Bay, Canada
22 Des Plaines disturbance, Illinois
23 Duckwater Crater, Nev.
24 Dycus disturbance, Tennessee
25 Dzioua Craters, Algeria
26 Ellef Ringnes Island Craters, Canada
27 Eyre Peninsula Craters, Australia
28 Flynn Creek structure, Tennessee
29 Franktown Crater, Canada
30 Glasford structure, Illinois
31 Glover Bluff structure, Wisconsin
32 Gulf of St. Lawrence arc. Canada
33 Gwarkuh Crater, Iran
34 Hagens Fjord Craters, Greenland
35 Haviland Crater, Kans.
36 Henbury Craters, Australia
37 Hérault Craters, France
38 Holleford Crater, Canada

39 Howell structure, Tennessee
40 Hungarian Plain, Hungary-Romania
41 Ilumetsa Craters, Estonia
42 Jeptha Knob structure, Kentucky
43 Ka-imu-hoku, Hawaii
44 Kaalijärv Craters, Estonia
45 Kalkkop structure, South Africa
46 Keeley Lake, Canada
47 Kentland structure, Indiana
48 Kilmichael structure, Mississippi
49 Kofels site, Austria
50 Lac Couture, Canada
51 Lake Bosumtwi, Ghana
52 Lake Dellen, Sweden
53 Lake El'gytkhyn, U.S.S.R.
54 Lake Humeln, Sweden
55 Lake Michikamau, Canada
56 Lake Mien, Sweden
57 Lake Siljan, Sweden
58 Lonar Lake, India
59 Macamic Lake, Canada
60 Malha Crater, Sudan
61 Manicouagan-Mushalagan
   Lakes area, Canada
62 Manson structure, Iowa
63 Mecatina Crater, Canada
64 Melville Island Craters, Canada
65 Menihek Lake area, Canada
66 Merewether Crater, Canada
67 Merriwell Lake, Canada
68 Middlesboro Basin, Ky.
69 Morasko Craters, Poland
70 Mount Doreen Crater field, Australia
71 Murgab Craters, U.S.S.R.
72 Nastapoka Islands arc, Canada
73 Nebiewale Crater, Ghana
74 New Mexico Crater, N. Mex.
75 New Quebec Crater, Canada

76 Odessa Craters, Tex.
77 Panamint Crater, Calif.
78 Paris (Sucy-en-Brie and Alentours)
   lakes, France
79 Parry Sound Crater, Canada
80 Patomskii Crater, U.S.S.R.
81 Pilot Lake, Canada
82 Pretoria Salt Pan, South Africa
83 Richât Crater, Mauritania
84 Serpent Mound structure, Ohio
85 Sault au Cochons structure, Canada
86 Syan Crater, U.S.S.R.
87 Semsiyât dome, Mauritania
88 Serpent Mound structure, Ohio
89 Sierra Madera structure, Texas
90 Sikhote-Alin Craters, U.S.S.R.
91 Socotra Crater, Socotra
92 Steinheim Basin, Germany
93 Sudbury Basin, Canada
94 Talemzane Crater, Algeria
95 Temimchât-Ghallaman Crater,
   Mauritania
96 Tenoumer Crater, Mauritania
97 Tiffin Crater, Iowa
98 Tunguska event, U.S.S.R.
99 Tvären Bay, Sweden
100 Ungava Bay, Canada
101 Upheaval Dome, Utah
102 Versailles structure, Kentucky
103 Vredefort structure, South Africa
104 Wabar Craters, Saudi Arabia
105 Wells Creek area, Tennessee
106 West Hawk Lake, Canada
107 Wilbarger dome, Texas
108 Wilkes Land structure, Antarctica
109 Winkler Crater, Kans.
110 Wolf Creek Crater, Australia

Missiles from outer space

All is dark, and the night sky glitters with the scattering of an infinitude of stars. Suddenly, infinitesimally, the pattern is disturbed, and a star breaks loose. A meteor, enveloped in flames, approaches Earth at cosmic velocity and, in a gleaming shower of sparks, plunges finally into the Earth's crust.

The vision, once seen, is not to be forgotten. But the moment of arrival of these 'signs and portents' from the unimaginable reaches of space is rarely witnessed – only half of the 2000-odd meteoritic falls known to science have actually been observed, and of the tons of cosmic matter which bombard the earth each day very little is recovered.

To poet, philosopher and ordinary awed gazer at the skies, it is the meteor in flight which fires the imagination – this object 'originating in the air' — as the Greek word meteoros suggests. To scientist and technologist, of course, it is the captured meteorite that counts, and the natural space probe that it represents.

The true origin of meteorites is not yet known, but it has been held that they are fragments of a planet within our solar system, shattered by a cosmic catastrophe. Present-day concepts also indicate that some meteorites are as much as four billion years old – practically as old as the earth – which suggests that our solar system crystallized at more or less the same time. Controversy over the origin of meteorites is still rife, and it will take much effort and study before science can provide more precise information about these celestial visitors.

Man's recent explorations in space have stimulated a rapidly growing demand for meteoritic material, and the discovery of a new meteorite is now of considerable scientific importance. Space technologists and the needs of various research programs are using up much of the existing material, and there is a great scarcity of newly-found material. Recent 'falls' or those with interesting flight markings or other unique features are worth more than gold, so that successful meteorite hunters – often using war-surplus mine detectors – can find their activities to be very profitable.

In fact, meteorites have been paraded in courts during law suits arising from ownership disputes between finders and owners of property, and in some countries they are considered to be a national treasure belonging to the State.

At 9 a.m. on April 28, 1927 a young girl in Juashiki was hit by a meteorite which is now preserved in Kwasan Observatory. There are

Chandpur, India

An artist's representation of the falling meteor which terrorised villagers at
Chandpur in April 1885. A stony meteorite, weighing 1.1 kg, was the only
specimen recovered.

also many records of meteorites crashing through houses or damag-
ing cars, or of showers of meteoroids whizzing around people like
bullets. But so far there is no proof of anyone being killed by a
meteorite, nor has there been an announced destructive collision of
one with a space vehicle. Airplanes near the Earth's surface are
protected by the atmosphere which burns up most of the falling
bodies, but this protection only extends to a height of about fifty to a
hundred miles, and the effect of a hit on a space craft by even a small
meteoroid could be disastrous.

Some meteoroids roam the universe in the region of the fixed stars,
others orbit the sun, and still others are the remnants of comet tails
that have been captured by the gravitational pull of Jupiter, our
largest planet, and return periodically to the vicinity of Earth to be
seen as meteor showers. They give us the delight of a fantastic vision,
and hold the promise that one distant day Man will understand the
secrets of the Universe.

Folklore and history

From earliest times the records of mankind have been filled with references to objects that have fallen from the sky. When, in about 2,000 B.C., an iron meteorite fell in Phrygia, it was for long afterwards, in the hands of priests, regarded as an object of veneration and – according to Titus Livius – when subsequently transported to Rome, it was worshipped there for another five hundred years.

The Chinese recorded a fall over 2,500 years ago, and appearances of "stars in rapid motion with a bright train behind" have been noted in the writings of every ancient race. These visions of a brilliant fireball in the sky, the showers of rocks, and the flaming path of meteors turning night into day transfixed and terrified their observers. They were variously regarded as messengers from the Gods, omens to be interpreted by only the most gifted astronomers, or message-bearing gifts rained upon them by far-distant ancestors.

The Arabs worshipped meteorites, long before the coming of the Prophet, as manifestations of God, and some such stones became objects of particular adoration. The famous 'Hajar al Aswad', for instance, Black Stone of Ka'bah, in the holy city of Mecca (see page 39), is almost certainly a fallen meteorite. According to legend it was originally white and only became black through the sins of man.

The instinct to venerate these fragments from the sky has been common to all civilisations, east and west alike; and evidence of this abounds in India, China and Japan, in the carefully preserved tribal treasures of Australian aborigines, and in the ruined tombs of American Indians. (The Casas Grande meteorite discovered in the burial ground of the Montezuma Indians in Mexico was found to be wrapped like a mummy.)

To ancient civilisations comets and meteorites had a predominantly ominous connotation – they were harbingers of divine wrath. But by the Middle Ages of our own era they had come to be regarded rather as a sign of divine protection. A well-documented record of a meteorite which fell in the Alsatian village of Ensisheim in 1492 indicates that the Emperor Maximillian himself saw in it a sign of heavenly protection against the threat of Turkish invasion. The 127 kg meteorite was prudently chained to the church wall, and part of it remains in the village to this day.

The veneration of meteorites has, therefore, been a consistent reaction on the part of mankind. It has been linked, however, with the instinct to put their magically regarded properties to practical use. The first known metal was almost certainly meteoritic iron (iron itself is not found 'free' in nature), and evidence remains of the tools and weapons that were forged from it – arrowheads (see page 36), Eskimo knives, Arabian swords, Javanese daggers and the like, all believed to have supernatural properties which rendered their users invincible.

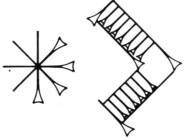

The earliest reference to meteoritic iron

Cuneiform characters impressed on a tablet, 3000 B.C.

Many rare artifacts worked in meteoritic iron have been recovered by archaeologists, and scholars have deciphered scripts and symbols describing meteorites and their falls. There was even a precise differentiation made – once the Egyptians began to import manufactured iron from Asia – between the 'iron from Heaven' and other metallic materials.

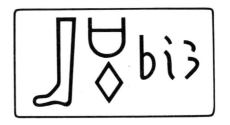

The Egyptian hieroglyphic symbol for meteoritic iron

Assyrian texts from around 1900 B.C. describe the activities of Anatolian traders, and reference is made to a commodity so precious that it had eight times the value of gold – a metal called KU.AN or Amatu, which evidence suggests was meteoric iron. In being worked into ornaments and weapons, it could be polished to a silver-like lustre; but perhaps its high market value lay principally in the magical properties so widely believed to derive from its heavenly origins.

Neo-Babylonian texts, for their part, reveal a less obviously materialistic approach but show that their soothsayers had cornered a lucrative market in interpreting these phenomena in terms of omens and medical diagnoses. Some of these predictions are intriguingly specific, for example:—

> "If a fireball flashes and it is flashing as bright as daylight, and it has a tail like a scorpion ... it is a favourable omen, not for the

master of the house, but for the whole land . . ." (Ref. Thompson, 1900, 200)

"If a meteor flashes and disappears in the middle of a star constellation: there will be a revolt". (Ref. Thompson, 1900, 237)

"If two shooting stars in Scorpio flash and move to the Western end: it will rain twice in that month."

"If, from above the Wagon Star, a meteor flashing like greenish lapis lazuli passes to the right of a man: that man will attain unto long life." (Ref. Langon, 1913-1923, 232-235)

"If somebody starts out on an undertaking, and a shooting star flashes from the right side of the man to his left side: favourable results." (Ref. Virolleaud, 1911, 116-125)

As on a clear night several meteors could be observed in an hour, it is not difficult to imagine that foretelling events in this manner could become a thriving business.

Although something is known of these age-old artifacts, there are many ancient scripts whose meanings are still not clear, and much treasure trove – from Mari, Kish, Tell Agrab, Khafajah, Warka, and other places – remains to be subjected to modern analytical techniques.

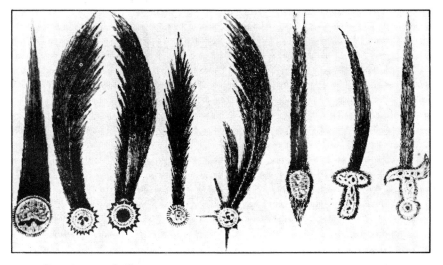

Cometary symbolism

Primitive people regarded comets as signs of their gods' anger, which required placating with prayer and sacrifice. Even in more recent times cometary symbolism expressed a sense of threat and foreboding, as shown in these sword representations dating from 1600.

The Earth under bombardment

It is impossible to establish with any degree of accuracy how much cosmic material actually reaches Earth. As interplanetary dust and micro-meteorites offer a minimum of resistance when passing through the atmosphere and vaporise readily, they cannot be observed or photographed, nor can the frequency of the falls be evaluated.

On a clear night, a keen observer, well posted, might spot as many as a hundred meteors, and even more if he used low-power binoculars. The greater magnification of telescopes would reveal to him even

Descent of a meteor

This engraving recorded the fall of a meteorite at Hurworth, near Durham, England, in October 1854.

Large cosmic wanderers — now called 'Earth-grazing asteroids' — are known to skim our atmosphere. The asteroid belt also contains tens of thousands of smaller bodies, and an incalculable reservoir of tiny fragments, all representing the material of an aerial bombardment.

more readily the tremendous amount of cosmic matter falling constantly into the Earth's blanket of air. It is fortunate for the survival of our species that the Earth does have this protection against cosmic debris, and that most of this material is burned up and vaporised before it reaches the surface. Otherwise we should feel like the victims of sporadic machine-gun fire directed at us from heaven.

The upper atmosphere, with its gas molecules, provides the tremendous friction that heats these cosmic missiles to incandescence, forming a luminosity that enables us to see a 'falling star'. It is a rare meteorite that successfully runs this gauntlet of air, and museums house a total of only some 500 tons of meteorites, representing 2000 different falls, either seen or later stumbled upon. Some were individual meteorites, others were offspring from a single body that broke up under the stress and.became a 'shower'.

As mentioned earlier, the total collected debris must be only a small proportion of all that arrives on earth, much of which falls into the oceans or the uninhabited areas of the world.

According to a hypothesis put forward by Swinnie and Petersson, meteorites are a recent phenomenum, perhaps occurring only during the past 25,000 years, and it is true that not many meteorites in collections have an Earth age any older than that. However, recent finds conclusively negate this theory. One of the objections to it is that it does not take into account the rapid oxidisation of iron. We know how quickly iron rusts – to-day you could kick a hole through a heavily-plated battleship of World War I. There are other objections to the Swinnie and Petersson theory, but it is quite possible that there were long periods when the Earth was either more or less subjected to bombardment than to-day. Perhaps there have been changes in the cosmic 'climate' analogous to the proven climatic changes on Earth.

The frictional destruction of meteorites creates incredible amounts of 'cosmic dust', which has been collected and identified beyond doubt by scientists such as Professor Maud Makemson. Tests with rainfall and snow have also been specially positive in showing the presence of nickel. Other researchers, carrying out more complex chemical tests, have sought to avoid areas where pollution was responsible for stirring up magnetic dust. Dust from areas where tests revealed the presence of coal particles had a different appearance and, although magnetic, contained no nickel (which is always present in meteorites).

Estimates have varied greatly over the years as to the weight of cosmic dust falling on Earth, and figures ranging from 5 to 50 grams per square kilometre have been quoted in the past. With improved methods of collection, these original estimates have now been increased by a factor of ten.

Calculations appear to show that the Earth is constantly growing heavier day by day. However, soil analysis reveals a remarkable absence of nickel, and no nickel means no meteorites. Throughout the past millions of years layers of cosmic matter could have been deposited to a depth of 10 to 20 centimetres, and one would think that some traces of nickel should be found. Admittedly nickel leaches out easily from decomposing meteorites, but this does not keep the total absence of it in the topsoil from being a mystery.

Occasionally, big masses of matter from outer space will survive their plunge through the atmosphere, only to disintegrate upon hitting the Earth. Over the aeons, our planet has been subjected to a few truly cataclysmic shocks, evidenced by the craters formed in the areas of impact, but this has not happened since Man arrived on the scene, and no one has witnessed any such events in recent times.

Nevertheless, certain rock formations continue to challenge geologists. There is, for example, what appears to be a meteor impact site some 50 kilometres wide located near Kentland, Indiana, only about 100 kilometres south of Chicago. One geologist, Robert Dietz, suggests that the meteorite struck so hard and with such velocity that it compressed the surrounding rocks instead of simply forcing them aside, thus bringing the deeply buried Ordovician rocks close to the surface. Yet no meteorites have ever been found in the Kentland area, and there are no signs of volcanic eruptions. Robert Boyer, another geologist, suggests that the "force was explosive, intense and sudden, and the shock came from above". The meteorite theory still seems to be the most probable, and there are many other craters which show this 'inverse stratigraphy'.

The answer to the mystery of many depressions widely distributed around the world has not yet been found. Take, for instance, the geological features characterising the Rieskessel Caldera in Bavaria, Germany, where there is a 'floor' measuring over 20 kilometres in diameter. Other circular depressions which have been discovered on a planet-wide basis, and subsequently proved to be craters of a meteoritic origin, were previously believed to have been produced by various local geological stresses and strains.

To meteoriticists, everything is a clue, and many ancient legends are strangely in accord with the latest scientific conclusions. For instance, the Hopi Indians say that one of the three gods that they recognise descended from upper space in fiery grandeur, forming the Diablo Crater Canyon in Arizona, used ever since for their major ceremonies. If the other two gods descended, we do not yet know where, but, acting on the hint of the legend, we have begun to search for them.

Stories of mystery and horror surrounding meteorite falls have come down to us through the ages. One report, dated 1135, from Prague, recounts the following: "A giant stone as big as a house fell out of the sky on the plain of Thuringia. People living in the vicinity could hear the noise it was making three days before-hand. As it hit the earth, half of the stone became embedded in the ground and for three days it remained red-hot, like steel when it is taken out of the fire". Elsewhere, from time to time, huge chunks of ice have been reported to have dropped from the sky, though these reports have not been proven.

A much more recent event, still being investigated, has only added to the mystery, and perhaps also to the horror. What can only be described as a cosmic bomb caused terrible devastation in the forests near the Tunguska River in Siberia (60°54'N, 101°57'E) on June 30, 1908, the nearest witnesses being about 50 kilometres away from the impact area. The light from the falling object was so intense that in far-away Sweden people could read the newspapers at 2 o'clock in the morning. In England, the evening hours became lighter instead of darker, and seismographs all over the world registered an earthquake traced to central Siberia. For a long time afterwards, there were beautiful sunsets throughout the world, red, green and yellow irridescences.

Years later, Professor L. Kulik of the Academy of Moscow, led several expeditions to the area to investigate this gigantic explosion, and he collected the accounts of a number of witnesses, despite the fact that at the time of the fall Russia was in a state of turmoil and World War I was raging. Vast distances had to be travelled over the tundra and it was enough of a problem for the investigating team simply to get to the site and return. In 1928, after a delay of twenty years, the first reports were finally published. A 'pillar of fire' 1,500 metres wide and 20 kilometres high rose into the sky. The forest was set ablaze over an area of 15 square kilometres and millions of trees were blasted out in a radial pattern, spreading over 50 kilometres. There was a column of smoke high in the sky shaped like a mushroom – a form that is now terrifyingly familiar to us but impossible to explain in 1908. A herd of 1500 reindeer browsing some ten kilometres from the impact area were blasted to ashes, and nothing could have survived within a radius of 30 kilometres. Although some witnesses claimed that they found pieces of 'bright iron' after the fall, Professor Kulik did not bring any samples back to Moscow.

This was the largest impact of modern times and it has been very thoroughly investigated since that first report. But the mystery only deepens and the debate goes on as to the origin of the impact. Some new evidence about this explosion, supporting the meteorite assumption, was introduced when iron oxide and silicate were found in the area.

Investigators like Dr Korobianikov reject the hypothesis of a giant solid meteorite. As an expert in shock waves, he suggests that sonic booms preceded the impact of the head of a small comet (generally composed of chunks of ice, solid matter and blocks of frozen gases), which was first sighted on an east-west trajectory some 600 kilometres north of Lake Baikal. Although the main mass has been estimated to weigh over a million tons,   only microscopic metallic spheriods have ever been found.

According to Dr Korobianikov, tree rings in the area show a characteristic formation, very similar to that caused by nuclear tests. Now, as cobalt is found in some iron-nickel meteorites, it is possible that radio-active cobalt was present in the head of the comet; and we can theorize that there was a tremendous amount of frozen ice and other material compressed in this cosmic body that arrived so catastrophically in Siberia in 1908. The low level of radio-activity in the area has even given rise to explanations that include black holes, anti-matter, and flying saucers.

Some theories hold that the impact of giant meteoroids on our planet could have released enough energy to shift the axis of rotation of the Earth. There is, in fact, real evidence that points in this direction, and many geophysicists accept the idea that some geographical structures, having diameters as large as 250 kilometres, could have been formed by meteorites when the earth was young. Explosion products like coesite, stichovite and suevite are criteria in identifying impact craters, since these are not produced by volcanic explosions.

The cosmic catastrophe theory affords an interesting explanation for the calendar reform in 238 B.C. Astonishingly accurate observation had already established a reliable working calendar, but this reform found it necessary to introduce the leap year and to fix the year at 365 days instead of 360. A possible explanation is that the impact of a large asteroid body sufficiently altered the Earth's previous orbit to necessitate the revision.

Mr M. J. S. Innes, of the Dominion Observatory of Ottawa, says: "There is a growing conviction that, during its early history, the Earth may have had a landscape similar to that of the Moon to-day ... Acceptance of meteoritic impact as a dominant factor in the development of the Earth's early surface may provide an explanation for many sedimentary basins which cannot be accounted for by erosion."

The geological features of the west coast of the great Hudson Bay, with a diameter of 450 kilometres, could suggest meteoritic origin. A. O. Kelly and F. Dachille drew attention to a huge submerged crater off the Carolina coast, with a diameter of 2,250 kilometres; its central dome reaching to the Bermuda Islands. Lunar craters of meteoritic origin of this size do exist. Of course, circular form alone is not enough proof of meteoritic origin and much more elaborate research is needed.

It is generally agreed that asteroid collisions could easily break off big chunks of matter, many kilometres in diameter, and if we liked to carry our pure speculations a bit further we could maintain that even ocean basins could be of meteoritic origin. A simple look at the atlas suggests a number of possible sites: the Gulf of Mexico (850 kilometres in diameter), the Sea of Japan (1,250 kilometres), and the Weddell Sea in Antarctica (with a diameter of about 2,000 kilometres).

It has been calculated that each year some 3,000 tons of cosmic debris fall to Earth in the form of dust, small meteorites, and occasionally much larger bodies. This figure when applied over geological eras would suggest that 15 trillion tons of the iron now on Earth are of extraterrestrial origin – a bombardment of truly awe-inspiring proportions.

How did meteorites originate?

"Galaxies are to astronomy what atoms are to physics".
Allan Sandage

The present scientific evidence indicates that meteorites are the most primitive samples of planetary matter which we can examine. Recent estimates, by isotope dating, show that some meteorites are 4,600 million years old – 600 million years older than the oldest Earth rocks – and suggest that they have come to Earth from the asteroidal belt between Mars and Jupiter. But the theories produced by astrophysicists and astronomers do not fully agree on the exact origins of meteorites, for which there are still numerous hypotheses.

The Swedish scientist, Svante Arrhenium, believed that stars eject matter which ultimately finds its way to Earth; about 150 years ago some scientists believed that meteorites were ejected from the Moon; others have theorized that volcanic eruption, in the past, ejected meteorites from the Earth up into outer space; and growth from gaseous interstellar matter has also been proposed as the origin of meteorites. But since Hladni suggested the extraterrestrial origin of meteorites 150 years ago no evidence has turned up against this view.

To-day many facts seem to indicate that meteorites are broken-off pieces from small planetary bodies in the asteroid belt, or pieces from comets that break off as they approach the Earth. If we imagine a cross-section of the Earth itself we would see three elementary components in a continuous series: the iron core; a stony-iron mixture; and the stone outside, the silicate shell. It can be deduced that if the Earth broke up, it would give three types of meteorites corresponding to the three different layers: irons, stony irons and stones.

It is generally assumed to-day that meteorites come from a part of the solar system. It is also believed that the body from which a meteorite sprang must originally have been part of a very large body similar in size to Earth, the gravitational field of a small body not being sufficiently strong to cause the separation between the iron and stone layers. Or it could be that one of several planets have broken up, thereby creating the bodies in the asteroid belt.

Despite the fantastic successes that men have had in exploring space, many complications arise when we delve deeper into the heart of the problem. Isotope measurements prove conclusively that meteorites are extraterrestrial bodies. We believe that they have come from the solar system, but we do not exclude the possibility that they might have come from interstellar space as a part of a comet. The origin and formation of meteorites will continue to be debated and, with new knowledge yet to be acquired, we shall get more and more definitive answers to help us to understand the immensely complex problems of the universe.

In February 1969, in Allende, Chihuahua, Mexico, a shower of stones fell after a bolide was seen, the stones being subsequently identified as carbonaceous chondrite (see page 38). One of the meteorites of this shower was dated to be 4,610 million years old, which probably represents the primordial planetary material. According to Dr Gerald Wasserburg of the California Institute of Technology, the Allende meteorite origin can be traced to the formation of the solar system, and the original gas cloud which formed it. It is the earliest thing condensed out of the solar nebula we have yet found. The majority of meteorites available for examination prove to be much younger and to come from different parent bodies that collided and broke up in the asteroid belt between Mars and Jupiter.

A few years ago a meteorite hit the moon close to the Apollo-14 landing site, and this was the first time that a meteorite impact had been recorded by one of the seismic 'listening posts' planted on the moon by our astronauts. It was estimated that the meteorite was three metres in diameter and that the crater it produced could easily have been the size of a football field – the shock being so great that the instruments had to be adjusted by radio signals sent from Earth. It is believed that this event will provide new evidence to clarify the theory of meteorite impacts.

Every new meteorite find throws some light on problems that are still obscure, but we are far from understanding fully the innumerable collisions among meteoroids in space. We still have no overall explanation of how the solar nebula came into existence, creating the planetary formation and the primitive meteoritic matter.

There is no evidence that meteorites fell before the middle of the Quaternary period – the youngest and present geologic period. Why this should be so is still the subject of continuing controversy.

The gigantic chemical laboratory of which we are all a part is being subjected to examination with passionate interest. We are likely to learn more about the origin of meteorites in the very near future, as space probes to Encke's and Halley's Comets are planned when they next approach the Earth. This will certainly bring more direct evidence as to the origin of the solar system and meteorites. The collision theory holds that planetary bodies collided and became part of the asteroid family in such a way that some of the matter may eventually have become meteorites; and this view is favoured by many cosmologists.

In fact, the mineralogical evidence concerning composition, structures and development of meteorites by slow cooling from a liquid state strongly suggests that the parent body was possibly somewhere between the Earth and Mars in size, and was created under a pressure of a million atmospheres and at a temperature of some 3,000°C. Orbits, velocity and other astronomical information about various bodies in space are all closely connected with the formation of meteorites and that of the solar system. Meteoriticists look forward to solving many puzzling problems about these 'cosmic missiles', in particular, by analysing recently observed falls; but they are still hoping to find a meteorite that has come from outside the solar system.

As theories on meteorites are modified and new solutions found, they merely open the way to new mysteries and new ways to explain the beginning – the big bang in which pure energy exploded. It is difficult to conceive these phenomena so far removed from anything we can actually experience. For instance, there are 'black holes' where the laws of physics do not appear to apply, and where energy and matter are sucked in, to disappear and somehow be pulled through a passage to emerge as 'white holes'. The Omega and Alpha – the ending and beginning – these are the forges of new universes. Whether the 'big bang' theory is true or not still remains in doubt; but the sophisticated monitoring of the space environment now being established must, in time, yield answers to these questions.

How meteorites reach the Earth

"Now slides the silent meteor on, and leaves
A shining furrow, as thy thought in me".

Tennyson

Meteoritic bodies travel in space in a spinning fashion, and generally in an orbit which is the same as that of Earth but with greater velocity. Meteorites that the Earth encounters 'head on' have a different speed from those overtaking the Earth from behind. It has been established by radar and radio observation that the speed of a meteorite entering the atmosphere can vary from 12 to 70 kilometres per second (43,000 to 250,000 km per hour).

On April 7, 1959, a network of camera stations was filming some very faint meteors when, suddenly, an entirely different, brilliant meteor appeared in the part of the sky under observation. The cameras were immediately switched to this new target, to record its passage until its ultimate impact in the area of the town of Pribram, near Prague, in Czechoslovakia.

The multiple-station recording of this Pribram meteor was so precise that not only was its point of impact predicted by computer, but from an analysis of its flight path it was possible to calculate that it had travelled in an elongated orbit originating in the asteroid belt between Mars and Jupiter.

This photographic survey in 1959 was the first time that the precise orbit of a meteoritic body had been calculated and the meteorite itself recovered. In all, nineteen stones (olivine-bronzite chondrite), having a total weight of 9.48 kg, were found, the largest fragment weighing 4.25 kg.

The majority of meteors can be seen at a height of some 100 kilometres. Usually fireballs (bright meteors) appear to have the size of the full moon and may be visible with their flaming trails down to 20 kilometres from the ground. The Pultusk meteorite (see page 77) that fell near Warsaw, Poland, and was seen by eye-witnesses while it was still at a height of 300 kilometres, was well observed and competently described. The colour, at some 180 kilometres, was bluish-green, but it turned red as it got lower. Although the size of the fireball was estimated to be about 300 metres in diameter, the largest specimen found of this meteorite shower was only 9 kg.

Fireballs are frequently described as being several hundred metres in diameter, with the size gradually diminishing as they near the earth until, at the moment of impact, the actual meteorite seldom weighs more than 12 kg. The colour of the path of a meteor is usually white, but can be reddish, yellow, or even green, but the glow of a meteor stops well above the stratosphere, at a height of about 12 km.

When investigating a meteorite fall, the really critical factor is the disappearance of the meteor trail in the sky, and the search-party will need to have as much information as possible about this in order to establish the point of impact. Compass bearings should be plotted to fix the trajectory, and eye-witnesses should be interviewed for information on the exact timing of the various phases, the appearance of the meteor, the duration of the burning period, the nature and timing of the sounds, and the current formation and direction of drift of the clouds.

Interplanetary matter, as it enters the atmosphere on a collision course with Earth, produces a succession of loud cracks and sounds very similar to sonic booms. Many blasts follow one after another which sometimes sound like rockets and can even make the ground tremble. In fact, the entry of a meteorite can be quite terrifying, as the following report illustrates:

Sioux County, Nebraska, USA, 42°35'N, 103°40'W.
Fell 1933, August 8, 10.30 hrs.
Stone. Eucrite (pyroxene-plagioclase achondrite)
20 stones fell, the largest weighing 4.1 kg.
As the meteor appeared in the sky, it grew in size, becoming more and more brilliant. There was a deafening roar which terrified people and livestock in the area, and the tremors set up by the shock waves gave the impression that an earthquake was in progress.

Fortunately for us, most interplanetary matter entering the atmosphere is destroyed while still at a great height, and broken down into particles of finest dust; the proportion that survives the passage and lands on Earth as meteorites is very small indeed. Meteoroids, which are also believed to be part of the cometary debris, usually seem to be of very low density and very fragile. There are others of high density, and the difference between the two can be established upon entry into the atmosphere. Many people around the world were very frightened when it was announced that Sky-Lab was doomed and would plummet to Earth. Subsequently, the author examined a spherical nitrogen container from Sky-Lab, which had been recovered from the debris that fell in Australia. It measured 75 cm in diameter, giving a surface of 1.7 square metres, and was marked by the impact scars of hundreds of micro-meteorites which, though very small in size, had penetrated quite deeply into the skin.

It is generally believed that a meteorite is very hot after falling through the atmosphere. This is not necessarily the case, and meteorites found in moist ground have even been covered with ice, showing their low internal temperature. Space itself is extremely cold and so is meteoritic material – with a temperature as low as −250°C

– before entering the atmosphere. During its short, fiery passage through the atmosphere the outer layers of a meteorite are melted away, but as the heat penetrates only a few millimetres below the surface, the inside remains as cold as ever. Iron meteorites are heated to a greater depth than stone ones because they are better heat conductors.

Most meteorites that enter the atmosphere burn up and are completely consumed within a second or two. After seeing a meteor blazing its way across the sky, it is easy to see how the notion arose that it would be hot, even molten, when it landed. But we know that in 1938, in Benld, Illinois, USA, a meteorite pierced an automobile, shredding the upholstery, without leaving the slightest trace of a burn (see page 63). Nevertheless, reports keep coming in of finds that are too hot to handle. Although it may be hard to verify reports like these, it is equally hard to deny them.

There are less than ten observed falls each year but, although opinions differ widely, the estimated total number of falls is about 500 – many of which must occur in oceans, deserts, on mountains, and in uninhabited places. Meteorites seldom bury themselves more deeply than one metre, and usually much less. Some even protrude above the surface, their impact speed having been checked so much by the friction of the air during their passage through the atmosphere.

Why study meteorites?

For ages past men have been fascinated – without precisely knowing why – by these fragments falling from the sky. Present-day science is no less fascinated by the unique opportunities they provide to study, in the hand, phenomena not to be found on earth. A new discipline has been born – the science of meteoritics – and many branches of older disciplines, among them physics, chemistry, metallurgy and biology, find in this new study a tool of incalculable significance.

– For chemists, the chemical compositions of meteorites enable them to study conditions that cannot be created in any laboratory, and which include a remarkable abundance of 'rare earth' elements. The noble metal content, too, is higher than in the Earth's crust – which was one reason for the prolonged but vain search for giant meteorites. Also, petroleum-like matter has been found in meteorites indicating that they were not over-heated during their flight through the atmosphere.

– To metallurgists, the structure of meteorites is of especial interest. Created under exceptional heat and pressure, some of the structures are unlike anything on Earth's surface, especially the intergrowth pattern of the minerals. This can be seen when iron meteorites are sawn across, a slice ground and polished to a very high lustre, and then etched with a dilute solution of nitric acid. These studies permit simulation of the space environment, a new technology that is developing rapidly as an inevitable consequence of our exploration of space.

– Nuclear physicists are particularly interested in the radio-activity (too weak to be harmful) in meteorites. Cosmic rays in space leave an 'imprint' on meteors, thus providing pointers to their age and 'curriculum vitae'. This is of prime importance because cosmic rays are harmful to human tissue, and meteorites contain information that will help Man to immunise himself against such radiation in his forthcoming travels through space.

– For biologists, molecules of complex carbons and amino-acids (the basic constituents of protein) found in certain meteorites provide fascinating evidence of life in space. The presence of organic matter in such meteorites indicates that they have come from other planets where there is life.

– Aerodynamicists, too, have something to learn, and a study is being made of the effects of the supersonic flight of meteors through space. Meteorites bear the distinctive marks caused by the streaming of air over their surface, and with the knowledge thus acquired improvements can be made in the design of slower-moving spacecraft and rockets.

Caroline Lucretia Herschel (1750-1848)

This brilliant astronomer, sister of the more famous Sir William Herschel, made an incalculable contribution to his work on the motion of the solar system in space. With the small, mobile, Newtonian telescope with which she constantly searched the heavens, she discovered three nebulae in 1783, and eight comets between the years 1786 and 1797. It is space wanderers such as these that approach earth and shower us with meteoric debris.

Caroline Herschel helped her brother to construct the great, reflecting telescope with which he discovered Uranus, and helped to augment their income by grinding and polishing hundreds of specula for other people's telescopes.

She died at the age of 98, outliving by twenty-six years her brother who had, strangely enough, originally come to England from Germany to earn his living as a musician.

What are they made of and how are they classified?

All meteorites have particular characteristics which permit their identification and classification. Almost all contain metals, particularly iron and nickel, whereas these metals are rarely found in their pure state on Earth. This, then, is the first clue to identification. In addition, they include minerals which are unknown on Earth, although they contain no elements that have not been discovered in the Earth's crust.

Meteorites are classified into three major groups, generally called Irons, Stones and Stony Irons, although sometimes other names are attributed to them.

Iron meteorites are also called metallic meteorites, iron-nickel meteorites or siderites. Aerolite is a term often used for stones or stony meteorites, which means that there is an absence of metal or that, at most, 25% of the meteorite is of metal content (mainly nickel-iron alloys). For stony irons, the term siderolites is also used. The chemical constitution of meteorites shows an abundance of elements, which include nearly all those that are known.

IRONS
It is estimated that iron meteorites make up some 10 per cent of all meteorites discovered. They are coarsely crystallised and appear to have cooled for millions of years in large bodies of sizeable asteroids, indicating that they had an insulating layer at one time.

They are often pitted and grooved as a result of their burning passage through the atmosphere. The direction in which the specimen points as it falls can be determined, and this is called 'orientation'. Upon impact, the surface is bluish-black, with traces of metal melted during the flight. After exposure on earth this group of meteorites rusts brown, and an oxide scale can be found on the surface, this thin layer being magnetite (Fe_3O_4).

The iron meteorite group is rich in metallic nickel-iron, and the principal accompanying minerals are troilite (iron sulfide), schreibersite (iron-nickel-cobalt phosphate), graphite (native carbon) and cohenite (iron-nickel-cobalt carbide).

Iron meteorites are divided into three sub-groups, according to their structure and to the proportion of nickel they contain: hexahedrites, octahedrites and ataxites. *(continued on page 41)*

How to catch a falling star.

The excitement of possibly finding their own fallen star makes children a wonderfully responsive audience for a meteoriticist.

Meteorite courses are now given in some universities and in a very few high schools, but a more widespread knowledge among adults – especially in areas where fireballs have been sighted – would lead to the more frequent recovery of valuable fresh material.

The Donato Comet

This painting confirms the general belief that the Donato Comet, pictured above Florence in 1858, was the most beautiful of the century. One could only hope to see a comet as bright as this once in a life-time.

Despite their brilliant appearance, comets have a comparatively small mass – a nucleus of only a few kilometres in diameter – but the tail may be over 100 million kilometres long.

The Donato Comet narrowly escaped destruction from wandering too close to the sun, and there is always the possible disaster inherent in a collision between a comet and our own planet.

Two characteristic specimens

Left:
A polished slice taken from the Brenham, Kansas, meteorite. the cavities in the iron-nickel frame are filled with olivine grains.

Right:
A weathered meteorite from the same fall, but in which the metallic reticulum has disappeared, leaving only a few translucent olivine crystals in the interior of the specimen.

This is a very good example of the effects of terrestrialisation.

Cross-section of an iron meteorite

This clearly illustrates the distinctive network of stripes crossing one another at specific angles – called the Widmanstätten pattern after its discoverer – which is characteristic of iron-nickel meteorites. This cross-hatched pattern which emerges when the specimen is cut, polished and etched with a weak solution of nitric acid is conclusive proof of the stone's meteoritic origin.

Swiss Alpine Fall

The brilliant beauty of a falling meteor, dramatised still more by a magnificent mountain setting, overawes the local inhabitants in this painting by R. Kierner.

Spotting from a plane

A bird's-eye view from a small aeroplane is ideal for evaluating the terrain of a probable impact site and the problems likely to arise when searching it.

Aerial surveys are also the very best means of identifying possible meteorite craters for subsequent examination.

Meteoritic iron arrowhead

This fine example of an arrowhead shaped from an iron-nickel meteorite is in the author's collection.

Iron-nickel meteorites have provided material since pre-historic times for the making of weapons and tools. Even at the beginning of the present century it was not unknown for blacksmiths in remote areas to use them.

A Tektite gem

This is a characteristic green Moldavite tektite from Czechoslovakia (4 × 4cm in size). Used first as jewellery by paleolithic man 25,000 years ago, Moldavites have been prized for their beauty ever since, sometimes set in gold or silver in their natural state, and sometimes cut into faceted gemstones.

Moldavites have the chemical composition and physical characteristics of glass, but a remarkable resistance to heat makes them more difficult to fuse than Pyrex glass.

Their wrinkled surface indicates the terrific speed of their descent through the atmosphere, and it is this aspect of the tektite mystery which interests scientists concerned with the re-entry problems of space craft.

Tektites found all over the world are similar in chemical composition, but their shape and colour show substantial differences.

La Madonna di Foligno

The date of Raphael's wonderful painting – 1511 – exactly coincides with the fall of the Crema fireball. If this had been blazing until it reached extinction about 30km from the ground, it would have been visible for over 600 kilometres. So, although he was 300 km away at the time, it is quite possible that Raphael actually observed the event which was incorporated in this picture, thought to have been commissioned by Sigismonde dei Conti to commemorate his escape from the meteorite.

Dei Conti is depicted as the red-robed kneeling figure, supported by St. Jerome, while St. John the Baptist is on the left, behind the kneeling St. Francis of Assisi. The fireball, with its glowing trail, is just above the cherub's head.

Mecca's sacred meteorite

This black stone – the 'Hajar al Aswad' – encased in a silver shell, is part of the Ka'bah which stands in the centre of the huge courtyard of Mecca's holy Mosque. Towards this a quarter of the world's population turns in prayer each day, and in pilgrimage once in a life-time.

A successful treasure hunt

The hunt usually begins with a survey made in a small aeroplane, from which a map is drawn of the area to be prospected.

In this particular expedition to recover iron meteorites, the first phase of ground activity was to locate nickel-rich metallic spheroids, using a simple magnetic rake. Where there was a positive reaction, holes were dug to test the soil for iron oxide. The area was then scanned with a device capable of detecting deeply-buried meteoritic material, and this ultimately revealed the 'find' illustrated in this photograph.

From outer space to the Empty Quarter

A lone rock in a desert region is always a possible meteorite. This iron-nickel meteorite, weighing 2,000 kg is the largest yet found in Saudi Arabia.

Hexahedrites

usually contain not more than 4 to 6 per cent nickel and have a classification symbol H. They consist entirely of kamacite (a nickel-poor, iron-nickel alloy). When a slice of this type of meteorite is cut and polished to high lustre and the polished surface is etched with a dilute solution of nitric acid, a rectangular pattern of fine lines appears, called 'Neumann lines' after their discoverer. Hexahedrite meteorites can be cleaved in three perpendicular directions along the faces of a hexahedron – hence their name. There are some 60 known hexahedrite meteorites, of which only 10 per cent were observed to fall, the others being found later.

Octahedrites

contain 6 to 12 per cent nickel. With the increased nickel content, this type of meteorite has an additional mineral called nickel-rich taenite (as well as nickel-poor kamacite). The two minerals, kamacite and taenite, are oriented to each other parallel to the octahedral plane, and hence their name. The arrangement of kamacite and taenite in octahedrite meteorites shows up as triangular bands parallel to the sides of an octahedron, and is brought out by etching the polished surface with dilute nitric acid. This structure is called the Widmanstätten pattern (see page 34), after the Austrian observer Baron Widmanstätten. These figures, whether fine or coarse, can be very beautiful. This unusual design shows lines of the crystal structure: it is unique to meteorites and is proof of meteoritic origin. The distinctive network of stripes crossing one another at certain angles is characteristic of a great number of metallic meteorites.

A sub-division of octahedrite meteorites is based on structures having kamacite bands of different thicknesses: the coarsest structures have bands 2.5mm thick or more; coarse have bands of 1.5 to 2.0mm; medium 0.5 to 1mm; fine 0.2 to 0.4mm; and the finest are 0.2mm and under. There are known to be more than 400 octahedrite meteorites – the Diablo meteorite from Arizona is a prime example – of which only some 30 were seen to fall.

Ataxites

are nickel-rich, with more than 12 per cent nickel. When etched, this group does not show a Widmanstätten pattern, nor can Neumann lines be seen, so they are difficult to distinguish from similar terrestrial structures. The intergrowth of kamacite (mineral of low nickel content) and taenite (which is much richer in nickel – up to 50 per cent) in ataxites is called plessite. This structural element, plessite, fills the spaces between the lamellae and can be observed with the naked eye. It is not an alloy but a fine mixture of the two minerals. There are some 60 known ataxite meteorites, of which only one or two have actually been observed to fall.

STONES

This is the most abundant group of meteorites, sub-divided into two major groups according to their structure: chondrites and achondrites. Stony meteorites are composed of various non-metallic minerals, usually silicates, and have only a small amount of metal in them, which is helpful for identification purposes. They are not pitted like irons, as they do not melt easily, but are rounded and have a thin fusion crust which is often crackled. There have been over a thousand stone meteorite falls and showers, and more than half of them were seen to fall and have been recovered.

Chondrites

contain small rounded grains of olivine and pyroxene (mainly hypersthene or diopside) in a finely grained aggregate, and taenite and kamacite may also be present (see page 78). The rounded grains are called chondrules, from the Greek words chondros for grain, and may vary from microscopic size to about the size of a pea. On examination under a microscope (using thin section technique) a radiating structure is often seen.

Chondrites have been divided into five sub-groups, according to their mineral content: enstatite, olivine-bronzite, olivine-hypersthene, olivine-pigeonite, and carbonaceous, each group having particular diagnostic features. Some chondrites contain hydrocarbons and amino acids, and black veins are a prominent characteristic of the majority of them. Some scientists believe that the organic matter found in certain chondrites is of biological origin.

Achondrites

are stone meteorites which do not contain chondrules, and are rarely found because they so much resemble ordinary terrestrial rocks. They have an igneous or brecciated texture, their metal content is very low, and they are more coarsely crystallised than chondrites. The Novo-Urei meteorite which fell in Russia in 1886 is an achondrite, and was the first meteorite to be found which contained diamonds. Achondrites differ in composition and texture from chondrites and have little or no free iron.

There are some 70 known achondrites, of which only about a dozen have been seen to fall and were recovered afterwards. Achondrites, which show similarities to certain terrestrial rocks, are divided into two groups: calcium-rich and calcium-poor, and there are three sub-groups of calcium-rich and four of calcium-poor achondrites.

STONY IRONS

These meteorites are transitional types between the two main groups of Irons and Stones, having some similarities with both. The proportion of non-metallic and metallic constituents varies. Stony irons are sub-divided into four groups according to the silicate minerals associated with iron-nickel, the four sub-divisions being pallasites, mesosiderites, siderophyres and londranites.

Pallasites

are named after Simon Pallas (1741-1811), a naturalist and explorer, who found one in Krasnojarsk, Russia, in 1772. In this type of stony iron meteorite cavities in the nickel-iron are filled with a sponge-like network of crystals of olivine (representing an emulsion structure), which can be rounded or angular, the proportion of metal and silicate being about the same.

When cut and polished the light-coloured areas in pallasites are nickel-iron metal, whilst the dark-coloured parts are olivine, this being the mineral most often found in meteorites after nickel and iron. In pallasites this is particularly evident where the grains show crystal facets, and the cellular network of nickel-iron filled with olivine forms a very beautiful structure.

Mesosiderites

are aggregates of a nickel-iron alloy and silicate minerals of irregular size. There are only some 30 mesosiderites known – about half the number of pallasites.

Siderophyres

consist of nickel-iron enclosing bronzite and tridymite minerals.

Londranites

are named after the town of Londran in Pakistan, where this type was first found, and are aggregates of olivine and bronzite with nickel-iron as principal minerals. Only three examples are known.

Certain meteorites contain minerals not found elsewhere in nature, such as:—

– Daubreélite	$FeCr_2S_4$
– Farringtonite	$Mg_3(PO_4)_2$
– Lawrencite	$FeCl_2$
– Niningerite	$(MaFe)S$
– Oldhamite	CaS
– Osbornite	TiN
– Perryite	Ni_3Si
– Schreibersite	$(Fe, Ni)_3P$
– Sinoite	SiN_2O

Nickel-iron is the most common constituent of meteorites. Next is olivine (magnesium-iron silicate). Cobalt is quite common. Quartz is minimal except in one meteorite. Gold and copper have been found in some meteorites, but in very small quantities. Platinum and other noble metals are also occasionally found, and minute diamond crystals have also been detected in a few meteorites.

The classification of meteorites has become more precise as research has advanced. Here is the classification according to Mason:

IRONS
A. Hexahedrites
B. Octahedrites
    1. Coarsest
    2. Coarse
    3. Medium
    4. Fine
    5. Finest
C. Nickel-rich Ataxites

STONES
Chondrites
    A. Enstatite
    B. Olivine-bronzite
    C. Olivine-hypersthene
    D. Carbonaceous
Achondrites
A. Calcium-poor
    1. Enstatite (aubrites)
    2. Hypersthene (diogenites)
    3. Olivine (chassignites)
    4. Olivine-pigeonite (ureilites)
B. Calcium-rich
    1. Augite (angrites)
    2. Diopside-olivine (nakhlites)
    3. Pyroxene-plagioclase
        a) Eucrites
        b) Howardites

STONY IRONS
    A. Olivine (pallasites)
    B. Bronzite-tridymite (siderophyres)
    C. Bronzite-olivine (londranites)
    D. Pyroxene-plagioclase (mesosiderites)

Current techniques classify meteorites by petrographic and genetic groups as well as by other more specialised systems.

Tektites

Tektites, it must be said, are not true meteorites, even though ancient civilisations venerated them equally as magical objects from the sky, and subsequent scientific thought tended to agree with the association.

We do not yet know the precise origin of tektites (whose name comes from the Greek word Tektos for molten), though fresh theories frequently enliven international symposiums on the subject. They have been variously claimed to originate on the moon, to be the remains of comets which exploded high in the Earth's atmosphere, or to be solidified droplets of molten rock splashed into the atmosphere by the impact of giant meteors on Earth. The theories continue, and so does the controversy.

What we do know, however, is that tektites, although they bear unmistakable signs of a flight through space, have – unlike meteorites – spent relatively little time outside the Earth's atmosphere.

Tektites do not contain the isotope ^{26}Al which is produced by cosmic radiation, and measurements of cosmogenetic isotopes show that they cannot have been formed at a distance from Earth greater than that of the moon.

Their basic chemical make-up is unique, though it bears a resemblance to obsidian, or natural glass. It consists mainly of silica, but the internal flow structure of particular inclusions differs from those in natural glass.

Tektites are amorphous – therefore singly refracting optically – and have a perfect conchoidal fracture and a pronounced vitreous lustre. Their hardness, on the Mohs scale, is about 5½, their specific gravity about 2.35, and they contain numerous microscopic air bubbles.

The shape of tektites derives from their high-velocity entry into the Earth's atmosphere, and their surfaces are pitted and grooved. They are to be found in many forms – spheroid, ovoid, pear-shaped, bulbous, lenses, dumbells, spindles, besides many of less regular shape.

In colour they are usually black, dark-brown or brownish-green, but a pale yellow variety, often of large size, is found in the Libyan Desert, and a green and often perfectly transparent variety in Moldavia (see page 36). These latter were found in strata of the late Tertiary and early Quaternary periods and contain some 80 per cent silica.

The largest known tektite was found in Saravane Province, Laos, in 1932. It weighs 3,150 grams, and now reposes in a Paris museum.

Although tektites have fallen to earth at different times and throughout the world, they are only to be found in certain areas and are then named after their site of discovery. Thus –

'Australites' from Australia and Tasmania
'Rizalites' from Rizal Province in the Philippines
'Ivory Coast tektites' from West Africa
'Malaysianites' from Malaya
'Billitonites' from Billiton Island, Indonesia
'Bediasites' from Texas, U.S.A.
'Indochinites' from South West Asia, Laos, Thailand, Cambodia, and North and South Vietnam
'Moldavites' from Bohemia and Moravia

Tektites from Thailand

Both the tektites illustrated are from the author's own collection: the disc-shaped specimen on the right is 8cm in diameter, 2cm thick and weighs 165 grams; the elongated stone on the left measures 9.5cm by 3cm.

Tektites are not true meteorites, although their grooved and pitted form is the result of a searing voyage through the atmosphere. The continuing mystery of their origin still tantalises scientists.

Geologically speaking, tektites are comparatively recent objects – Bediasites being some 35 million years old, Moldavites 15 million, Ivory Coast 1½ million, and Australites a mere 3,000-5,000 years old.

For centuries tektites have been used as gemstones and, in the Far East, as talismans, and references to them go back as far as 2,500 B.C. In Sanskrit their name is Agni Mani, which means 'fire gem'; the Chinese term is 'Huo Chu', or 'fire pearl'; and they have often been referred to as 'chips from the stars'.

Moldavites – named after the Moldau (Vltava) river in Bohemia, where the first tektites were found – have been used as gems to a much larger extent than any other tektites. Specimens of exceptional beauty and symmetrical surface have been on the market for over 100 years, and have been used in the fashioning of some magnificent jewellery in Czechoslovakia.

These Moldavian tektites are generally darker in colour and there are some specimens which are perfectly patterned and display a beautiful, green, glittering surface in transmitted light. They have been used for faceted gems and cabuchon cuts, and their naturally baroque shapes are very popular.

The key to the tektite mystery has yet to be found. Although we know that these curious objects, with their fascinating surface configuration, are close relatives of meteorites, their fall has never yet been observed. However, their composition and form are still being evaluated in chemical and aerodynamic terms, and these studies are making a significant contribution to lunar research and to the solution of the re-entry problem of space vehicles.

So, for the moment, we must limit our conclusions to the fact that tektites are of an extraterrestrial nature, but have arrived in our atmosphere without having made a long journey through the Solar System.

How to identify true meteorites

Not long ago, a mis-shapen lump of iron in use as a barn door-stop on a Canadian farm turned out to be a meteorite, and more have been found embedded in buildings and other unlikely places. Such cases are rare, but many fragments are still to be found which were once meteors streaking through the night sky with luminous brilliance. Although they may be found anywhere on earth, searches would bring meagre results unless a recent fall had been observed in a well-defined area. However, any stone with peculiar markings could, on examination, prove to be a meteorite.

As there are rocks of different shapes, sizes, colours and textures all around us, it is obviously not easy to identify a meteorite amongst them. Yet there are certain clues which, once learned, become engraved upon the memory.

The Innisfree Meteorite

This meteorite, located by the Canadian Meteorite Camera network, belongs to the hypersthene chondrite stony group. Its black fusion crust is about .03mm thick. The upper scale shown in the illustration is in inches, the lower one in centimetres, and the specimen weighs 2.07 kg.

Iron meteorites are the easiest to identify; they are heavy, look like iron, are strongly magnetic, and are covered with pits and depressions (see page 35). If one has fallen recently, its surface will still have a black crust, but with time this will wear off leaving the surface with a rusty appearance. Some have unusual shapes but many, at first glance, look just like ordinary, common rock or minerals. Their characteristic surface pitting may be either deep or shallow, but is seldom jagged. A visit to the meteorite collection of a museum will be of great help in recognising cosmic origin.

The presence of nickel in addition to the iron is an almost sure proof and can be detected in a laboratory in the following manner. One gram or less of the specimen is dissolved in dilute hydrochloric acid; the solution is then heated to boiling point and a few drops of nitric acid added together with a drop of citric acid. The solution is then cooled, carefully neutralised with ammonia, and a few drops of dimethylglyoxine solution in alcohol added. If nickel is present, a scarlet-red deposit will appear. There are many other tests for nickel, but if this one is positive, further investigations have to be made, particularly with regard to structure.

To do this, a slice is cut off from the metal to be investigated. It is then ground to a plane surface on coarse carborundum, and subsequently ground with successively fine grits – up to 400 or more. The final polish is given with a fine-grained polishing material like aluminium oxide, or felt with tin oxide; at which point the polished slice should show a high lustre.

Once the specimen has been thus prepared for etching, a 6 per cent solution of HNO_3 (nitric acid) is brushed over the slice. This process, which takes only a few minutes, should make what is called the Widmanstätten pattern visible. The appearance of an octahedral structure on an etched iron-nickel alloy is definite proof of its meteoric origin.

The octahedral structure found in certain iron-nickel meteorites can also be brought out by the slow heating of a polished surface. The two iron-nickel alloys (kamacite and taenite) will oxidise at different speeds and create a lovely pattern. Octahedral structure is a complicated process of crystallisation and can be observed when the nickel content is between 6 per cent and 12 per cent, and the distinction is made between octahedrites of coarse, medium or fine structural components of kamacite bands of a thickness of 2.5mm or less. When a slice of octahedrite is heated to about 900°C (melting point is much higher), the octahedral lamellae disappear and cannot be restored.

But not all iron-nickel meteorites reveal such striking structural patterns as octahedrites. The hexahedrites, which contain a smaller proportion of nickel (4-6%), do not show the Widmanstätten pattern but show instead a series of parallel lines called Neumann lines.

Whilst the ataxite group show neither of these structures on etching and so are harder to identify. Both these groups were described in greater detail in a previous chapter.

Stony meteorites are the most difficult of all to identify. They often have irregular shapes and shallow surface pits, and their thin, black fusion crust – formed by the heat of passage through our atmosphere – is sometimes broken off in certain areas to reveal a lighter colour. Some have a smooth fusion crust and can be of angular shape; a chipped surface can show different colours inside of black, brown or grey.

Specks of metal are a good indication that the stone in question in indeed a meteorite, as such specks are extremely rare in earth rocks, and a polished slice will frequently show such metallic grains. A magnet can also reveal the presence of iron. So, any rock displaying flakes of metal should always be investigated.

One identifying feature of stony meteorites is small spheres – often visible to the naked eye – of silicate minerals on broken or polished surfaces. These round or oval grains, called chondrules, are found associated with silvery specks of iron-nickel.

Meteorites seldom have sharp corners because of friction during their passage through our atmosphere. They are usually irregular in shape and each one is highly individual. Most have travelled in a tumbling fashion and so do not have a symmetrical form, but if they have travelled without tumbling they can be cone-shaped. The characteristic pitting of meteorites (which looks like finger-tip marks in clay) is believed to be formed during the last moments of flight.

Magnetic testing for meteorites is very helpful, but not conclusive as there are also some exceptional earth minerals that are magnetically responsive. If a crushed fragment of a specimen proves to be responsive, further laboratory tests should be undertaken.

Many minerals, ròcks and even man-made objects can be mistaken for meteorites (see page 77). When dark-coloured rocks are ground on an emery wheel to detect specks of white metal, it is easy to confuse metallic grains with glittering crystals such as quartz, pyrite, mica, etc., but if a magnifying glass is used, no mistake should be made. However, the list of possibilities for error is embarrassingly long; furnace slugs, certain volcanic rocks, some pyrite concretions with black surface crusts, oddly shaped pebbles and boulders – even heavy, iron grinding balls have given rise to confusion.

Although the distinguishing features of meteorites may be of such a nature that very advanced laboratory tests are required, the majority of them have sufficient characteristics for a trained eye to detect them with a minimum of equipment.

Here are a few basic rules to help eliminate non-meteoritic rocks:—

Fossils and shells do not appear in meteorites.

Formations in layers do not appear in meteorites.

Meteorites are not round, well-shaped spheres.

Crystals of mica or quartz are not visible in meteorites.

Meteorites do not contain bubble holes.

Meteorites are usually heavy. Even stony meteorites are usually twice as heavy as granite.

The fusion crust of meteorites is of a very peculiar nature and has sometimes been compared with the black crust of bread.

A simple test is to rub a corner of the suspected rock with emery paper (for best results use medium grit). Grind to a depth well below the surface and be on the look-out for small irregular grains of shiny, silvery metal. If such grains are visible the specimen should be further investigated with more advanced laboratory techniques.

Advanced laboratory tests for meteoritic identification can be carried out in most mineralogical laboratories. Many institutions and museums are particularly active in meteoritics, especially where they are linked with a vigorous programme of space exploration.

Dating a meteorite

Meteoritics is a young science confronted with the problem of identifying and analysing very old material – material much older than the Earth itself. Moreover, the problem is a double one: that of determining the cosmic age of meteorites and also their terrestrial age – that is, the length of time they have been on Earth.

Any attempt to estimate the age of a meteorite requires a broad understanding of the universe and star systems beyond our own, and their original formation. When a meteorite is examined, there are five basic factors to be studied:—

  i) the process of the forming of the parent body from the elements;
  ii) the separation of the iron from the silicates, due to the heating of the parent body;
  iii) the cooling of the parent body;
  iv) fragmentation of the parent body;
  v) the fall of the meteorite.

Although everything in space is being continuously bombarded by cosmic rays, these can only penetrate about a metre; only the outer layers of a body, therefore, are affected, whilst the material deep inside remains untouched. But once the parent body of a meteorite breaks down into smaller fragments of less than a few metres in size, cosmic rays are absorbed by the fragment and a radio-active process of decay sets in. Since we know the rate of the decay processes, we can thus calculate the cosmic age. After the fragment lands on Earth, it is protected from further cosmic radiation by the atmosphere that serves as a shield; so by measuring the number of cosmogenic isotopes contained in a meteorite, the period of time during which it was subjected to cosmic rays can be established.

The principle of radio-active decay – called the 'radio-active clock' – is also used to establish the terrestrial age of a meteorite. Age determination is based on the essential characteristic of certain elements to break down with time into other elements. One is radium – the decay product of uranium (the parent element), which decays further into the element radon, which decays still further. Such complex laboratory experiments indicate that certain meteorites are older than the oldest rocks on Earth.

During drilling operations to locate oil in Zaputa County, Texas, a meteorite was reported to be found in 1930, at a depth of 165 metres – in the stratum of the Meocene age. Based on stratigraphic measurements, it was reckoned to be 40 million years old. If proved, this would make it the oldest meteorite on record.

So far, we have not discovered any concentrations of iron, nickel, cobalt and other elements in the soil which could lead us to assume that they had originated from huge decomposed meteorites.

Isolating isotopes in meteorites is very difficult, as sometimes the proportion is only one part in three million. Age studies of meteorites have led scientists to the conclusion that the solar system and meteorites are in the order of 4.5 billion years old (Scientific American, April 1957). Different isotopes in many meteorites have been under study by various research workers, and all have reached roughly the same conclusion. As chemical activities within meteorites permit a dating technique which is constantly improving, it should now be possible to recognise traces of meteorites which fell on earth in early geological times.

Falls and finds

A description of a meteorite always states whether it is a 'fall' or a 'find'. By definition a 'fall' is a meteorite that has actually been seen to fall and has been recovered afterwards at the place of impact. A 'find' is a meteorite, not observed to fall, but found and identified as such.

A fireball above Madrid

A luminous meteor pictured in the sky above Madrid. It fell on February 10th, 1896, creating terror among the inhabitants. Calculations based on the interval between the flash at the time of explosion and the subsequent detonation noises suggest that the meteor disintegrated at an altitude of 30 kilometres.

Stony meteorites resemble many terrestrial rocks found on the Earth's crust and are not usually particularly noticeable. Once the black surface crust has worn off with time, it is most unlikely that anyone would pick one up and identify it as a meteorite unless it happened to have some unusual features or appeared to be incongruous to its surroundings – such as in a desert or on a bed of ice.

Iron meteorites resist weathering effects better than the stony, and have a weight, appearance and shape that characterise them, so that they are therefore much easier to recognise. Arid regions and climates are particularly favourable for protecting the meteorites from rust and the terrain makes them easier to spot. A 'find' may be discovered after many thousands of years and yet still retain its cosmic identity. Smaller meteorites, however, 'weather' faster by exposure to the atmosphere; with time, their chemical composition is changed and their meteoritic characteristics lost. This is called the process of terrestrialization.

About two thousand meteorites have been recorded and analysed, and new meteorites are added to this list at a rate of about a dozen a year. This includes both 'falls' and 'finds', and with increasing interest in meteorites and denser populations, we can expect an increasing number of recoveries.

Obviously there are more 'finds' than 'falls'. Of meteorites recovered from 'falls', the great majority are stony, while only about 20 per cent are in the iron-nickel group. However, in the 'finds' category there are about twice as many iron-nickel as stony meteorites because they are the more distinctive, and so more likely to attract the attention of passers-by.

The largest stony meteorite ever recovered in the U.S.A. fell near Norton, Kansas, on February 18, 1948, and it was observed to burst twice during its descent, at estimated altitudes of 40 and 18 kilometres. The biggest stone from this meteorite shower, from which over 100 stones have been recovered, weighed close to one ton.

Since more stony meteorites than irons have been seen to fall, it is assumed that they must predominate in nature. Stony irons, which are a combination of the two, are extremely rare, and only about 80 have been recovered – of which a mere dozen were 'falls'.

It has been observed that more meteorites hit the earth in the afternoon than in the morning. In view of this, the current theory is that most meteoroids travel in the same direction around the sun as does the earth. Most 'falls', not unnaturally, are observed in the clearer months of summer.

Considering the immense quantities of coal that have been dug over the last centuries, it is very surprising that mining operations have not uncovered meteorites imbedded in coal which took such a long time to form.

In the dwellings of stone-age man tektites have been frequently found, but no meteorites. But when the Mediterranean civilisation was at its peak 'celestial' metal was often referred to in contemporary writings, suggesting that there must have been plenty of it around. Then, as centuries passed, there became less and less mention of 'celestial' metal, and during the period of the mechanical revolution, when all sciences leapt forward, very few meteorites were found.

These observations could indicate that the earth has been bombarded by meteorites more frequently at some times than at others, but human history is, however, relatively brief, and it would be dangerous to draw conclusions from a statistical survey that is so short compared to the time-scale of the phenomena themselves. Yet the absence of meteorites in mining operations, particularly of coal, must lead naturally to speculations that at some ancient time no meteorites fell to Earth. And even to-day, although many Earth-based cameras are scanning the skies over millions of square kilometres, surprisingly few meteorites are being found.

Observed 'falls' are much more valuable to science than 'finds' picked up later. For the meteoriticist, in addition to having a fresh and uncontaminated meteorite for examination, it is important for him to know, whenever possible, in which direction the meteor travelled across the sky, its type of smoke trail, colour, and sound effects, and the exact time and duration of the observation. A discovery of a meteorite is an extraordinary event and a careful record should be made as to the size of the crater, and the type of material in which the meteorite was found. The temperature of the meteorite should be estimated if it is found immediately after the fall, as, although it is commonly believed that all meteorites are hot upon impact, some of them may actually be frosted.

Names and meteorites

Meteorites are named by the location of their fall – the nearest inhabited place, a prominent nearby landmark, or a topographical feature such as a mountain. However, in order to avoid misunderstandings – occasionally the same meteorite is known by different names, and sometimes two or more bear the same name – precise information as to longitude and latitude has to be recorded.

Frequently more than one meteorite is found within a given area, as a result of the break-up of a huge meteorite as it lands on earth. In such a case the same name is given to all the specimens resulting from that particular 'fall', or 'find'. For example, the meteorite shower seen to fall in Tenham, Queensland, Australia, in the spring of 1879, produced 102 stony meteorites scattered over an area of some 25×5 km but all the stones belonging to this 'fall' are catalogued as Tenham meteorites.

A code has been devised by Dr Frederick C. Leonard in order to avoid possible confusion among names of meteorites, and systems and instructions for borderline cases for the naming of new meteorites are constantly being improved. A Nomenclature Committee of the Meteoritical Society has been set up in order to regulate the naming of new falls and finds, and new meteorite names should not be published without their approval. After approval, the name will be correctly placed in catalogues and no confusion will be possible.

If the name of the place of fall has, in the course of time, been changed, it has been suggested that the name originally given should, nevertheless, be retained. Another problem can arise from different spellings of the place of find or of very old names. Therefore synonyms for each meteorite are also given in the catalogue. Details of the fall or find which are recorded in the catalogue are: a description, chemical analysis and other information (including the literature on the particular meteorite), and the whereabouts of known specimens, particularly of the main masses.

Most recent catalogues of meteorites use a new classification scheme which is based on chemistry and texture, although so far it has only been adopted for newly examined meteorites. A world listing is due to be published under the auspices of the Commission on Meteorites of the International Union of Geological Sciences. Examples of the listing for three different meteorites are given below.

Bjurböle Borga, Nyland, Finland 60°24′N., 25°48′E.
    Fell 1899, March 12, 2230 hrs.

Stone. Spherical olivine-hypersthene chondrite.
One stone fell through the sea ice and broke into fragments, the
largest of which weighed 80 kg, the total weight being about
330 kg. Described and analyzed by W. Ramsay and L. H.
Borgstrom (Bull. Comm. Geol. Finland, 1902, no 12, p. 1). The
largest piece in Helsinki, the next largest in Stockholm. 798
grams in New York. Etc.

Stauton Augusta County, Virginia, USA.   38°9'N., 79°4'W.
Found 1858-59.
Synonyms: Augusta County; Folersville; Louisa County.
Iron. Medium octahedrite.
A mass of about 68 kg, found in 1858 or 1859, was thrown away
and later built into a wall, and in 1877 was taken out and
recognized as a meteorite. Analyzed by J. R. Santos, Ni = 7.5%.
6 kg in Vienna (Naturhist. Mus.).

Mühlau Innsbruck, Tyrol, Austria.   47°17'N., 11°25'E.
Found about 1877.
Synonym: Mercherburg.
Stone. Spherical chondrite, grey chondrite.
A stone of 5 grams was found, probably soon after the fall. (A.
Brezina, Ann. Naturhist. Hofmus., 1887, vol. 2, p. 115). Main
mass in Vienna (Naturhist. Mus.). This is the smallest meteorite
that constitutes an entire fall.

The Meteoritical Society, NASA Johnson Space Center, Houston,
Texas and the Center for Meteorite Studies, Arizona State
University, Tempe, issue the Journal of the Meteoritical Society
'Meteoritics' in which new meteorites are described, and topics of
interest to the meteoriticist discussed.

Sizes and showers

The largest known meteorite (one of the iron-nickel ataxite group), weighing 60 tons, was found in 1920 at Hoba Farm, near Groot-fontein, Namibia (South West Africa). It measures $2.95 \times 2.84 \times 1.25$ to 0.55 metres, and was found embedded in the earth to a depth of 1.5 metres. The original intention was to extract its metal content – 50 tons of iron, 10 tons of nickel, and ½ ton of cobalt, which would represent a small fortune to-day – but the South African Government forbade this and declared the meteorite a national monument, to be preserved where it landed.

The largest on display in any museum is the 34 ton meteorite named the 'Cape York' which can be seen at the Hayden Planetarium in New York. It was found and appropriated in 1897 by the Commodore Peary Expedition on the west coast of Greenland, and was nicknamed 'The Tent' or 'Ahnighito'. There were two other meteorites from the same fall, one called 'The Woman', which weighs 3 tons, and the other called 'The Dog', about ½ ton, all three being owned by the American Museum of Natural History. Knives made of iron from these meteorites, with bone handles, were given to Capt. John Ross in 1818 by the Eskimos of Prince Regent's Bay.

A fourth large meteorite from the same area was found in 1913 by the Eskimos on the Savik Peninsula, and was eventually removed to Copenhagen in 1925. The Eskimos considered these Cape York meteorites as a heaven-sent source of metal from which to make their tools.

The largest known stony meteorite was recovered after a meteorite shower in Kirin, China, on 8 March, 1976, the biggest fragment weighing 1750 kg. This fall was the most widely observed in history, as it occurred near the city of Kirin, which has a population of over a million, and many thousands of witnesses were subsequently inter-viewed.

Of the many meteorites that fall to earth each year, only a very few are found, and the chances of actually seeing a meteorite fall are, alas, minimal. As often as not, a meteorite will break up before impact with the Earth, and the fragments will be spread out over an area of several square kilometres. The usual size of a recovered meteorite is several kilograms, but there are much smaller ones – some so small that they cannot even be detected. In fact, meteorite dust can sometimes be identified in melted glacier ice.

When a large number of meteoritic bodies arrive at the same time the phenomenon is called a 'meteorite shower'. For a long time the only showers observed were composed of stony meteorites, which descend in an ellipsis in the direction of the flight. The relative air resistance of a large falling meteorite is less than that of a small one, so the larger meteorites in a shower will travel further. The result of this can be seen when recovering meteorites, as the size of the meteorites found increases along the direction of their flight. Stony meteorites found some distance apart from each other have been fitted together, like a puzzle.

The first observed shower of iron meteorites fell in a forest in the Sikhote-Alin mountains of eastern Siberia, on February 12, 1947. Over 100 small craters were produced and an even larger number of impact holes, meteorites being discovered both inside and outside the craters. The largest crater measured 28 metres across and the heaviest mass found was 1745 kg.

Showers of iron meteorites have been numerous but seldom witnessed. A typical example is the Henbury meteorite in central Australia, where 13 craters were found in an area of less than one square kilometre. The largest crater is 180 metres in diameter and the smallest 8 metres, and the heaviest meteorite recovered from this find is reported to weigh about 150 kg. A considerable quantity of small iron-nickel meteorites scattered over the whole region were recovered with special metal detectors after systematic search. Several of these craters hold rain water, permitting vegetation to grow in an area that is very arid, thus becoming easily detectable, particularly from the air.

Another well-known find with numerous masses recovered is the one known as the Gibeon (Bethany), South West Africa meteorite. This is another example of an iron meteorite shower, as is the one found in Coahuila, Mexico, which is of the hexahedrite group.

Very large meteorites weighing over 100 tons cannot possibly be found, as they will retain enough of their cosmic velocity when striking earth for the impact to vaporise them. They will usually form a crater, scattering their fragments around the area of impact.

Recent meteorite showers witnessed and well-documented have been observed in the following areas:
  L'Aigle, France – April 26, 1803
  Stannern, Czechoslovakia – May 22, 1808
  Knyahinya, USSR – June 9, 1866
  Pultusk, Poland – January 1, 1868
  Hessle, Sweden – January 1, 1869
  Khairpur, India – September 23, 1873
  Homestead, Iowa – February 12, 1875
  Mocs, Rumania – February 3, 1882
  Holbrook, Arizona – July 19, 1912
  Sikhote-Alin, Siberia – February 12, 1947
  Kirin, China – March 8, 1976

The total weight of meteorites recovered from the first nine showers listed above is just over one ton, and represents thousands of individual specimens, whilst the Sikhote-Alin, Siberia fall produced metal fragments weighing in total some 70 tons. The fall area of these meteorite showers varies from 1 × 2 km for the Sikhote-Alin shower, to 4.5 × 25 km for the one in Khaipur, India.

The average weight of meteorites recovered from more than 400 observed falls is 20 kilograms, but it has been calculated that the average extraterrestrial particle weighs only about one millionth of a gram and drifts to earth in the form of the finest dust.

Not surprisingly, over 70% of meteorites probably fall in the oceans which cover two-thirds of the globe. It is highly unlikely that topographical features of the earth such as mountains would have an effect on the distribution of meteorites, but, interestingly, almost half of all meteorites recovered, numbering nearly 1000, come from North America. This figure represents different types of meteorites, each having its own characteristics; but individual meteorites are counted as one, regardless of the number of fragments found, when they come from the same parent body that penetrated the atmosphere as a single event.

The Great Plains area in the United States is generally stoneless and densely cultivated, and strange rocks found there have often proved to be meteorites. Dr Harvey Nininger, who is said to be indirectly responsible for one-tenth of all meteorites recorded to date, has trained thousands of ranchers and farmers to be on the look-out for unusual stones, and he has regularly visited them to make identifications. His 'cosmic connections' have kept him in good health and he re-married in 1980 at the age of 92!

Many areas of the world have physical characteristics which make it easier to spot meteorites. So, if local residents could be trained to recognise them in the fields, we would undoubtedly discover many more meteorites.

A dangerous kind of beauty?

The flashing beauty of a 'shooting star' is a far-distant delight – indeed, its existence in the far reaches of space is a part of its attraction for us. We do not stop to think what that flash of beauty may possibly mean in human terms. We now know that our planet Earth is constantly bombarded by extraterrestrial bodies, and that some of them have been of great size. Assumptions have been made by scientists that the upper several metres of the Earth's surface are thickly mixed with weathered and decomposed meteorites fallen over geological ages. But, although specialists using modern dating instruments (radio-active 'clocks') have identified large craters which were formed by falling meteorites in historical times, we have no positive evidence of meteorites causing significant loss of life.

Nevertheless, there is evidence enough that people have from time to time had unnervingly close encounters. Take, for example, Mrs Hodges, of Sylacauga, Alabama, rudely awakened from a nap by a stony meteorite falling through the roof of her house in November, 1954. It was fortunately small – no more than 1.6 kg in weight – and she escaped with a severe bruising. The guests staying in a hotel in Caernarvonshire, Wales, in September, 1949, were even more fortunate. A stony meteorite weighing 794.05 grams fell through the

Meteorite starts a fire?

On January 16th 1846, a fireball was reported to have fallen and set alight a building in Chaux, Saône-et-Loire, France. This depiction of its magnificent train down to the point of impact owes more to the artist's imagination than to fact. We now know that meteorites are not incandescent during the final stages of their fall, but this particular one could well have caused the subsequent fire by knocking over a lamp or stove.

roof of the building, but no one was hurt. It was a black crystalline olivine-bronzite chondrite, subsequently named Beddgelert, after its location.

On a sunny morning in June, 1938, in Pantar in the Philippines, the local padre was holding Mass when his voice was drowned by what sounded like a series of explosions and thunderous vibrations that caused doors and windows to rattle throughout the town. One photographer got a clear snapshot of the meteor at the instant of explosion, and fifteen kilometres away people working in the fields saw fiery objects with tails of smoke and smaller objects 'as big as corn and rice grains' falling in a great shower. In spite of this spectacular display, only sixteen small stones finally ended up in museum collections to commemorate a fall which was almost unique in the richness of its phenomena and in the drama of its impact. It was described in such wonderful detail by so many observers that it has become a classic for meteoritic study. The fall was named Pantar, Lanao, Philippine Islands, and the stone was a veined olivine-bronzite.

At nine o'clock on the morning of September 29, 1938, a fragment weighing 1.7 kg fell through the roof of a frame garage in Blend, Illinois, U.S.A., and carried on through the roof of the family car. But nothing was burned, since meteorites are not fire hazards, and the complete stone is to-day in the Chicago Field Museum of Natural History.

A direct hit

On September 29th 1938 at Blend, Macoupin County, Illinois, a 1.7 kg stony meteorite fell through the roof of a car. It was described by a witness as "sounding like an airplane going into a power dive".

There is a record of a young girl being hit by a meteorite – in Juashiki, Japan, in the year 1927 – but fortunately the stone, now preserved in the Kwasan Observatory, was too tiny to harm her.

On January 18, 1916, a meteorite fell through the roof of a house in Baxter, Missouri, and the local newspaper (with the resounding name of the Stone County Oracle) published a description of its fall. The writer clearly sounds the note of alarmed bewilderment which the citizens of Baxter must have felt. "About 9 o'clock a.m. on Tuesday, the 18th, the citizens were shocked by a loud noise in the elements that seemed to begin right over us. Three loud reports in quick succession resembled blasts of several sticks of dynamite discharged. Then the noise changed to something like someone beating on a large boiler, and it travelled toward the north as it died away in the distance. Its duration was as much as a minute. Some say they heard it for fifteen minutes. It's got us puzzled, Readers, tell us if you heard it. Someone tell us your opinion of what it was. Some folks here thought it an airship sailing over, and that it turned loose some dynamite. It was cloudy, but the clouds were thin. It was heard eight miles south and six miles north, that we know of, and they all tell the same story about the noise and direction. We will be glad to hear from others in regard to it". Well, it was not an airship, but a meteorite, and it finally fell into the attic of a house in Baxter, doing no human damage. Subsequently named the Baxter, Missouri, meteorite, it was a chondrite, with numerous bright metallic grains embedded in a compact greenish matrix.

Meteorite craters

Every ultimate fact is only the first of a new series.
Ralph Waldo Emerson

The Arizona Crater is, by a long way, the largest in the world so far proved to be of meteoritic origin. This vast, circular depression – 1250 metres wide by 190 metres deep – was caused by the impact of an iron mass of some 2 million tons, the force of the explosion being equivalent to that of an atomic bomb in the 5½ megaton range. In a split second the meteorite removed some 200 millions tons of rock – or, to put it in perspective, about a quarter of the total amount excavated in the construction of the Panama Canal.

A cosmic bomb crater in Arizona

20,000 years ago a falling meteorite of enormous size blasted this crater — 190 metres deep, and 1250 metres in diameter (making a potential arena that could accommodate 3,000,000 spectactors). We know now that the main body of the meteorite was destroyed by the colossal impact, leaving only a widely-scattered shower of fragments.

According to carbon-14 and other recent dating tests, the meteorite struck the earth about 20,000 years ago, but the crater was only officially discovered by white explorers in 1879. Since then it has been a source of continuing scientific interest to meteoriticists, astronomers, geologists, physicists and ballistic experts – to say nothing of the million visitors a year who come just to gaze at the awe-inspiring sight.

At one time it was believed that the crater had a volcanic origin – a 'steam blow-out' caused by subterranean heat; and when, less than a century ago, it was first suggested that it had been caused by a meteorite, the idea was scoffed at as being fantastic.

In about 1906, a mining engineer, Daniel Barringer, noted that the iron fragments strewn around the crater had a 7% nickel content. He secured mining rights and thereafter spent most of his life drilling holes and shafts to reach the main mass of metal. Quicksands often hampered this work, and despite new drilling techniques, no sooner was one problem solved than another and more formidable one would take its place. He dug down almost 500 metres before he accepted that he was fighting a losing battle and abandoned the operation shortly before his death in 1929, after almost a quarter of a century of frustration.

The mass of metal has never been recovered, although its immense potential value (about $500 million at to-day's prices) has attracted many prospectors.

Towards the end of Barringer's life a new theory was propounded: that the meteorite had not simply fallen from the sky to Earth, but had grounded after a nearly level flight, as if descending from orbit. The theory goes that the descending meteorite pushed before it an immense mass of air heated to several thousand degrees and, upon striking the Earth finally, it exploded, sending up a gigantic column of pulverized matter. The impact would have been of such force that a part of the meteorite could have tunnelled deep under the rim of the crater, or the main body could simply have bounced off again, perhaps ending up in the ocean.

To-day we know that the impacted and the target materials must have disintegrated into fine particles and literally splashed out into the atmosphere. Fragments of the meteorite were scattered over a ten kilometre area, of which the largest weighed nearly a ton, with about two dozen pieces of over 100 kg. These specimens are now in museums or in private collections. The hunt for the missing meteorite body that could have been buried under the rim halted many years ago and will probably not now be resumed, as the area has become a National Park. In any case, it is most unlikely that the main mass still exists.

Existing fragments from this nickel-iron meteorite have the following composition: 90% iron; about 7% nickel; with traces of platinum, cobalt, chromium, copper and diamond. The two minerals, cohenite 3 $(Fe,Ni,Co)C$, and schreibersite 3 $(Fe,Ni,Co)P$ are abundant, neither of which is found in terrestrial rocks. Sulphides of iron and nickel are also present in the form of rounded nodules. Such metallic spheroids are often found on the side of meteorite craters and are droplets that have condensed from the cloud resulting from the explosion attending the impact of a giant meteorite. The meteorites found around the crater are in the iron division of the coarse octahedrite type. They were given the name Canyon Diablo, and the term 'impactite' has been coined to describe the melted rock fused with bits of meteoritic matter in the explosion.

There is strong evidence to support the meteoritic origin of many other craters. Over 30,000 have been observed on the hemisphere of the moon turned towards the Earth, of which several thousand are larger than the Arizona Crater. In fact, after the classification of the Arizona Crater, many others were added to the category as a result of the identification of meteoritic material in their vicinity.

Large, circular and elliptical depressions now act as an initial signal to geologists, for we know that when a cosmic body weighing over 100 tons strikes the earth its outer layer and a part of the Earth's surface vaporise in a terrific explosion that almost completely destroys the meteorite itself. Now that this blast phenomenon is understood and we can authenticate the meteorite fragments remaining, we have been able to add several new craters of meteoritic origin to our inventory; so far, only about a dozen such craters have been identified, but there must be many more waiting to be correctly classified.

Here is a list of meteorite craters that have been positively identified.

Aouelloul Adrar, Mauritania, 20°15′N, 12°41′E.
>250 metres across, 6 metres deep, found in 1920. No meteorite found, but only nickel oxide and silica glass.

Boxhole Plenty River, Central Australia, 22°37′S, 135°12′E.
>Recognized in 1937. 175 metres across and 10 to 15 metres deep. One meteorite of 30 kg was found and numerous smaller pieces in the gravelly alluvium. Numerous 'shale balls' (rounded masses with iron cores) have also been found there.

Campo del Cielo Gran Chaco Gualamba, Argentina, 27°28′S, 61°30′W.
>Recognized in 1933. Four shallow craters, the largest 70 metres in diameter and 5 metres deep. One big meteoritic mass was found by Don Rubin de Celis in 1783, but this area has a history of iron since 1576. Rock flour and silica glass have also been found.

Dalgaranga Western Australia, 27°45′S, 117°5′E.
>Discovered in 1923. The crater is 25 metres across and 5 metres deep. Signs of explosive action. Many small fragments of meteoritic iron were found.

Haviland Kiowa County, Kansas, USA 37°37′N, 99°6′W.
>Recognized in 1923. One crater 15 metres across and three metres deep. It was first thought to be a 'buffalo wallow'. This is

the site of the fall of the Brenham Pallasite meteorite (stony iron). Much unaltered metal and olivine were found there, the latter sometimes of true gem quality. The value of the material taken from the 'buffalo wallow' would be worth well over a million dollars today.

This meteorite was first noticed in the raw Kansas prairies near the end of the 19th century by a young bride, Mary Kimberly, soon after she came to the farm with her husband. Mary collected these strange rocks and in the beginning of the century Kimberly's farm became known as the 'meteorite farm'. The couple marketed their meteorites, and the event has entered into the folklore of the region.

Henbury McDonnell Ranges, Central Australia, 24°34'S, 133°10'E. Found 1931. Well defined group consists of 13 craters — the largest is oval, 200×110 metres and 15 metres deep. This is a well-authenticated meteorite crater which indicates an explosive origin. Iron meteorites were found as well as dark brown nickel-ferrous silica glass.

Kaalijarv Saaremaa, Estonia, 58°24'N, 22°40'E. Recognized in 1928. Of the six craters, the largest is about 100 metres in diameter, with a rim rising about 6 metres. Iron meteoritic fragments were found, after a long search, ranging from 0.1 to 24 grams, with a total weight of 100 grams. The total number of meteorites was 30, but the area has been populated for centuries, and larger pieces may well have been carried away.

Mount Darvin Tasmania. 42°15'S, 145°36'E. Recognized in 1933. One crater. Silica glass, 'Darvin glass of Tasmania' which contains coesite and germanium, and typical meteoritic crater glass were found. The crater from which the glass was expelled was probably destroyed by glacial erosion.

Odessa Ector County, Texas, USA, 31°43'N, 102°24'W. Recognized in 1929. Two craters, the larger being 160 metres in diameter and 6 metres in depth. Iron meteorites and 'shale balls' have been found around the crater. A meteorite museum has been erected next to the crater. This is a typical explosion crater chiefly identified by the efforts of the late D. Moreau Barringer, the mining engineer who bought the worthless land around the Arizona crater and formed the Standard Iron Company to mine the metal.

Podkamennaya Tunguska, Yeniseisk, Siberia, USSR, 60°54'N, 101°57'E. Formed June 30, 1908. Evidence points to a great explosion before impact, and the forest was destroyed within a radius of some 30 kilometres from the point of impact.

Sikhote-Alin Maritime Province, USSR, 46°9.6'N, 134°39.2'E.

Formed on February 12, 1947, and associated with a meteorite shower that made numerous craters and pits. The largest crater is 28 metres across. Many meteoritic iron fragments were found, the biggest some 300 kg in weight, and material was found scattered both inside and outside the craters. The four field parties from the Committee on Meteorites, USSR Academy of Sciences, reported that the bolide was so bright that it blinded those watching it, and that it cast moving shadows. The velocity in the upper atmosphere was calculated to be about 15 kilometres a second. Professor Krinov suggested that the meteorite exploded, in the atmosphere, into fragments weighing from a fraction of a gram to one block over a ton, and that it broke further upon striking the ground. The final count gave 122 craters and impact pits. Location of finds of this meteorite occurred in an elliptical pattern, smaller specimens apparently being more retarded by atmospheric resistance, and hitting the ground somewhat sooner. The meteor moved from north to south. The meteorites showed a chemical analysis of 93.5% iron, 5.2% nickel, 0.47% cobalt, 0.20% phosphorus, 0.06% sulphur and small quantities of other elements. This crater-making meteorite provided a rare opportunity for study and yielded much evidence leading to the correct identification of the origin of other craters.

Wabar Rub'al Khali, Arabia, 21°29'59"N, 50°28'20"E.

Recognized in 1932. Two craters: 100×100 and 55×40 metres across. Iron meteorites and silica glass were found. These craters are almost obliterated by drifting desert sand.

Wolf Creek Wyndham, Kimberley, Western Australia, 19°18'S, 127°46'E.

First observed from the air in 1947. Large circular crater, 900 metres across and 50 metres deep, the rim being of shattered sandstone rising 30 metres above the surroundings. This crater is next in size to the Arizona crater. Over 700 kilograms of oxidized meteorites were found, the largest weighing about 150 kg.

There are to-day about 60 important structures that are accepted to be of meteoritic origin. Although there are many more known craters in existence, in the absence of iron meteorites – either within the crater itself or in its vicinity – the judgement of authenticity is often withheld. However, in an effort to identify further impact sites, investigations are constantly being made of such geographical features as circular lakes, circular mountains with a ring-shaped ridge, oval lakes where the trees have been uprooted, and conical depressions. But in many of these it is hard to find meteoritic matter as a means of positive identification.

Prospecting for meteorites

"And all depends on keeping the eye steadily fixed upon the facts of nature and so receiving their images simply as they are".

Francis Bacon

Discovering a meteorite is a thrilling experience – indeed an unsurpassable one for the trained interpreter of its hidden messages. But, given the difficulties involved, the experience is also likely to be very rare, even for a skilled prospector. As well as a knowledge of meteorites, elementary geology and mineralogy are very important. Moreover, meteorites still awaiting discovery are, for the most part, in cold and arid regions where exploration is not easy, and transportation is frequently lacking. In fact, survival training is really a 'must' before adventures of this sort are undertaken.

Above all, though, lies the basic problem of where to search. In many cases, known meteoritic sites – such as the Arizona, Henbury and Odessa Craters – have been designated national parks, or otherwise put out of bounds to would-be prospectors, even though there is still meteoritic material to be found there. There are still known craters which are not forbidden areas, but they are becoming few, and are unlikely to be easily accessible. Catalogues categorise them as authenticated craters, doubtful craters, and impact holes, and excellent bibliographies exist for the guidance of hunters.

The chances are very small for a meteorite hunter to witness a fireball in the sky, and to be able to plot its path so as to predict the point of impact. He has less chance of stumbling over a meteorite by accident in the field, even if he is equipped to recognise it. One would really have to be born under a lucky star to find a meteorite! Nevertheless, attending an hour's lecture, a visit to a meteorite collection, or watching a relevant television programme would enormously improve his chances of a 'find'.

For the trained prospector, a good system is to follow-up with personal interviews the reliable reports of observers who have seen a fireball; then to set up surveying instruments in the prospect area and to work out the path of the meteorite from the information gathered. This method may be within the reach of only a few meteorite prospectors but, followed through with determination and patience, it can lead to rewarding discoveries.

'Productive' areas, once detected, are usually kept secret by their finders, who want to recover as much material as possible before others get there.

The basic instrument for collecting iron-meteorite specimens is a portable metal detector, often called a 'treasure-finder'. These instruments are usually sensitive to a depth of only half a metre, though meteorites buried to a depth of 4 metres and even more have been detected and recovered. For deeper exploration, much more complicated devices have to be used, and their cost is high. But for the most part meteorites, and especially the smaller ones, do not penetrate deeply into the ground, and erosion and the action of rain can even expose them to the air.

There are ingenious devices mounted on special sleds that can comb a considerable area in a relatively short time. Their detectors emit a shrill note to indicate the presence of buried metal, a needle registering a signal at the same time. Earphones are sometimes also included in the equipment. The prospector stops when the signal from the detector is strongest, and digs deep enough to recover his treasure – which may turn out to be an old horse-shoe, a rusted pipe, or, hopefully, a meteorite (see page 40).

Electromagnets are sometimes used, mounted on special vehicles, to collect iron meteorites on the surface or even, when the instruments are powerful enough, to recover them from a shallow depth.

A network of observation stations with ingenious electronic systems has been built in many countries, to photograph bright meteors and thus be able to calculate approximately where they fall. But the enormous cost of this operation, which recovers only one or two new meteorites a year, has already led to some segments of the network being abandoned. Hence the vital importance of independent prospectors, who can make a real contribution, based on their own enthusiasm, and at very little cost.

Such a one was Dr Nininger, who operated on his own for decades in the United States with the help of his family. He worked without any official financial support at a time when meteorite specimens were gathering dust in museum store-rooms. The major part of his collection was secured by the British Museum of Natural History and the Arizona State University at Tempe, U.S.A., for a relatively modest sum around 1960. A small part of the collection remained in the hands of younger members of the family and, with the accelerating programme of space exploration, it suddenly became of enormous scientific importance and monetary value. It was sold off recently to the Max Planck Institute in Germany.

In the early days Nininger himself bought meteorites from farmers who were glad to part with these oddities for a dollar a lb. weight. Today the value of some of them is $20 a gram, and this redoubtable pioneer, at the age of 93, has lived to see a thousand-fold increase in value for some specimens.

A successful meteorite prospector needs to have a good understanding of 'fall' phenomena and a knowledge of flight path calculation. Meteorite showers, which are of particular interest to search for, have sometimes been well-documented, and one of the main observations is that the meteorites found were scattered over an elliptical area, which can be more than 10 kilometres long. If the shower came down from a northerly direction, heading south, the smallest meteorites would be found at the northern end of the ellipse and the largest at the southern end, since with their proportionately smaller air-resistance heavy meteorites travel further than light ones. When records are examined, a good strategy is to note carefully the extent of the fall area, the number of fragments detected, and their total weight; then, to draw a map showing the distribution of recovered meteorites in the shower.

In recent years many meteorite craters have been identified and no doubt there will be further discoveries. Sometimes even casual observation from the air has led to the discovery of meteorite craters, and reconnaissance flights are very important in spotting geological features for further examination on the ground. Craters can be strongly weathered, and the crater field might have only a very slight resemblance to the expected configuration. Once on location, a systematic survey should be made and samples collected. Silica glass with iron-nickel inclusions on the surface of the ground is an indication of a possible meteoritic impact, but much laboratory work must be carried out to establish beyond a doubt the meteoritic nature of a crater.

Most craters of proven meteoritic origin are caused by the impact of iron meteorites. Doubtless gigantic holes have also been punched by stone meteorites, but these are much more difficult to identify because of the rapid terrestrialisation of the fallen stones scattered in the area. Verification of craters is made more difficult by the fact that large meteorites strike the earth with such great velocity that they explode and vaporise: this means that the meteoritic matter becomes very difficult to identify, and we are unlikely ever to find a single meteorite mass of more than 100 tons for this reason.

If crater structures are to be investigated thoroughly, aerial photographs should be taken of suspected impact sites, preferably from low-flying aircraft (see page 36). The author, for example, has used a Piper Warrior airplane for this purpose. Certain terrains make it particularly difficult to identify significant features – flat areas; or deserts, where characteristic circular shallow lakes or 'playas' fill with water and then evaporate, leaving mud flats. To recognise potential craters amongst these is not easy, and the landscape must be interpreted in detail and with critical care. Small, highly-manoeuvrable airplanes are required, not to mention a good photographer with a strong stomach! Aerial photographs provide a unique record, and the small airplane will continue to be of great help in the discovery of impact sites.

Prior study of photographs of known crater fields are valuable to those undertaking aerial expeditions because of the similarities
(continued on page 81)

Space-born diamonds

The nodule in the centre of this Canyon Diablo meteorite contains minute carbonado diamonds embedded in a matrix of graphite carbon.

Great heat and pressure are required to create diamonds from graphite and carbon; so that either the meteorite's impact produced the needed heat and pressure, or an earlier shock in its cosmic history did so. This latter theory is supported by the fact that several of the Canyon Diablo specimens show indications of earlier collisions, in which they were fragmented before entering the Earth's atmosphere.

"Moving about in worlds not realised"

The worlds in Morris R. Dollan's painting are not
those imagined by Wordsworth in his great ode
'Intimations of Immortality', but they are no
longer pure fantasy. The practical possibilities of
almost effortlessly towing great, mineral-loaded
asteroids into the Earth's gravitational field, and
giving them a controlled landing, are now being
actively considered.

A fireball entering Earth's atmosphere and showing an orange-coloured ionisation tail.

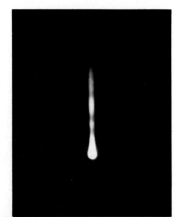

The same fireball disintegrating as it enters deeper into the atmosphere.

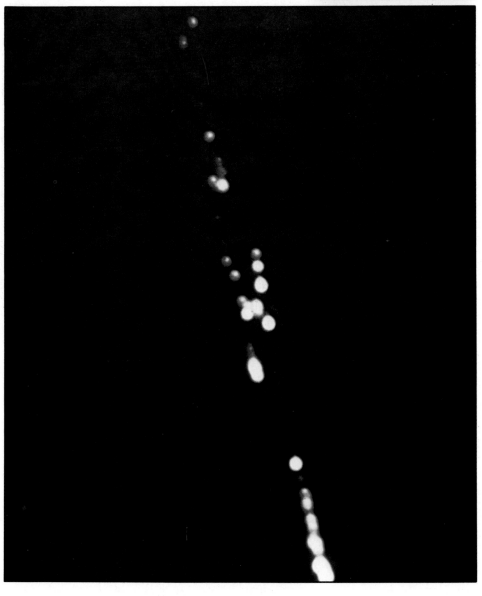

A cosmic masterpiece

These extraordinary specimens, now in the author's collection, are from the Murchison, Australia, shower which fell in September, 1969. The larger specimen, which weighs 80 grams and measures 5.5 × 4 × 3cm, has had its edges and corners rounded-off by its passage through our atmosphere. However, the smaller specimen has retained its basic angularity and has only a very thin fusion crust, which suggests that it is a surviving fragment from the inside of the fireball which was only subjected to friction after the main body had exploded.

This particular fall was remarkable for the wealth of information and new possibilities of research that it provided, including the first identification in a meteorite of amino acids (the basic building blocks of life) of extra-terrestrial origin. Belonging to the very rare carbonaceous chondrite group, the Murchison meteorite is estimated to have been formed in the solar system some 4.5 billion years ago – since when it has undergone but little change. It contained calcium, gypsum and spinel, rarely found in meteorites, and astrophysicists have related its composition to that of 'infallen' material found in lunar soil.

Spacecraft designers have also found it significant in their study of high-speed aero-dynamics and the problems encountered during re-entry into the Earth's atmosphere.

These are *not* meteorites

Furnace slag, pyrite or limonite concretions, lavas, basalt bombs, ferromanganese, ferrosilicon, lumps of iron, or even rusted iron objects may be mistaken for meteorites. The specimens illustrated here, for example, are:—

Left:
Olivine crystal from an 'olivine bomb' found in the Canary Islands.

Centre:
A commonly found pyrite nodule – this one from Champagne in France.

Right:
A manganite specimen from Odenwald, Germany.

Stone meteorite from Pultusk

This stony meteorite, now in the author's collection, fell at Pultusk, Poland, in January 1868. It weighs 41.5 grams, and measures 4 × 3cm.

The smooth, black, flight-marked crust is only shallowly-pitted.

The heaviest stone recovered from this fall weighed 9 kg, and the fragments were strewn over a typically elliptical area, some 8 × 1.5 kilometres.

Older than Earth itself

This carbonaceous chondrite, in the author's collection, fell at Allende, Mexico, in February 1969. It weighs 141 grams and measures $7 \times 5 \times 3$cm.

The cosmic age of the material in this fall is calculated to be 4,610 million years, making it probably the oldest sample we have of primordial planetary material – and perhaps even part of the supernova debris which predates the formation of the solar system.

The white inclusions in the Allende meteorite are thought to be some of the earliest objects to condense out of the solar nebula, and their isotopic anomalies are providing fresh insights into the origin of the solar system.

A witnessed fall

This white-veined, olivine-hypersthene chondrite, in the author's collection, is part of a shower which fell at Mocs, Roumania, in February 1882.

3,000 stones were recovered from this fall, scattered over an area of 12 square kilometres; one of them, weighing 36 kg having penetrated 60cm into the frozen ground.

Jewellery from space

Every piece of the rare and
beautiful meteoritic jewellery
created by the world-famous
Swiss designers Schlegel and
Plana is accompanied by
certification of the space
origin of its material.
Pendants, rings and
necklaces are fashioned from
small metallic meteorites, in
such a way as to preserve the
shapes and markings born of
their voyage through space.
Gold settings beautifully
recreate the incandescent
glow of the meteorite's
passage through the Earth's
atmosphere.

The Bennett Comet

Photographed by C. Nicollier from the Gornergrat Observatory, near Zermatt, Switzerland – March 1970.

between meteoritic craters. Even photographs taken by passengers in commercial aeroplanes have subsequently helped to identify previously unrecorded craters – several were reported in this way near Baghdad, others between Cairo and Basra, and there are other promising sites, where there has been no volcanic activity, which have not yet been investigated.

In the past, meteorites received little attention in the curricula of geology students (and even to-day this is the case in many universities). As a result, there are only a few people trained to identify the 'odd rocks' which are actually meteorites. But teaching people to become 'meteorite-minded', by giving lectures, distributing leaflets, radio, TV and press appeals, can do much to counteract the general level of ignorance. Boxes of old rocks and minerals can occasionally be found in attics and cellars – even in museums – which may have been lying neglected for years. They could contain a meteorite which, in all probability, had not been recognised as such and had been originally mislabelled by an amateur collector.

In the search for fresh finds there are general rules about where to look for meteorites and what to look for. For example:—

–   Any curious stone that appears in an unlikely place should be investigated (see page 40).
–   Rocks found on frozen lakes make excellent suspects and many meteorites have been found in such areas. This is because the fallen fragments were not heavy enough to break through the ice.
–   Sometimes a suspicious depression can be observed in a flat field and should be investigated, as well as any irregularities and holes that could have been made by falling objects.
–   Land cultivators often collect the rocks found in their fields and these provide a useful source for examination for fusion crusts.
–   Meteorite falls produce booming sounds like thunder. If the sky is clear and cloudless and no airplane is in the vicinity, such a boom may be caused by a falling meteorite, and merits investigation.
–   Rocks of peculiar shape and appearance should be examined carefully. Some meteorites are 'blunt-nosed'-orientated and such rocks, in particular, should be investigated. In fact the 'blunt-nosed' missiles of to-day have been designed with the help of ballistic and aerodynamic studies of meteorites.

The positive identification of a meteorite requires skill and experience. Sometimes it is wrongly classified as 'wind-blasted lava rock', 'magnetite', 'lava bomb', 'iron slag', etc. To avoid such errors, it is helpful for the student to study photographs showing the important surface features of meteorites, fusion encrustations, orientation, pitting or piezoglyphs, stages of terrestrialisation, various flight markings and typical meteorite fragments found around impact craters. Museums of natural history are the most likely places where such specimens and photographs can be found.

In the days when Canyon Diablo in Arizona was not a protected area, there was a cowboy there who could spot buried irons without a metal detector, by recognising faint concentrations of thin iron oxide on the surface; another claimed to have a horse who always pawed the ground over the right spot!

Lacking a meteorite-minded horse, there are other useful aids to successful prospecting! One is to establish friendly relations with the people living in an area where a fireball has been sighted. The great pioneer, Dr Nininger, has derived much help from this source, and has indeed trained many people to become meteorite-conscious. He has even found some specimens which farm-hands were using for weight lifting and once, when investigating a fireball in Wyoming, he was presented with a meteorite as a result of chatting to a petrol pump attendant. There must be many unreported meteorites still waiting to be recovered, some of them in the hands of people who are completely unaware that their strange stones are treasures.

Physical fitness is another necessity for the would-be hunter. A strong walker can search about 8 acres in one day – i.e. about one square kilometre in a month. Meteorite showers can be scattered over ten to a hundred square kilometres – so that the amount of leg-work involved can be very considerable.

Luck, too, plays an important part – as it does in most human activities. Dr Nininger himself found a meteorite quite by chance, while he was eating a snack at the side of the road in New Mexico, in May, 1944. Idly scanning the bare terrain, he noticed a small, dark pebble protruding from the ground. He loosened it from the soil and, in a few seconds, identified it as a meteorite. As he wrote afterwards, "Twenty years are not too long to wait for the thrill that it gave me to realise that here at last I had made a brand-new discovery, alone and unaided, in a region far removed from any known fall". The specimen weighed only 7.6 grams, had a fusion crust, and, with the aid of a magnifying glass, the chondritic structure with some grains of metal could be clearly seen. It was subsequently named Puento-Ladron, Socorro County, New Mexico. The excitement of this find by pure chance was quite different from that of patiently following up reports and eye-witness accounts of fireballs. The New Mexico area is well-suited for meteorite prospecting, and Dr Nininger has encouraged hunters to visit it. He believes that no stony meteorite entering the Earth's atmosphere reaches the Earth unbroken, and that all falls of stony meteorites are more or less a shower – so the area may well yield additional finds.

He also believes that future finds in this area should be larger – perhaps weighing several kilos – and that they are likely to be on the surface of the ground, so this desert area of New Mexico presents enticing possibilities for prospectors.

What to do if you see a fireball

A fireball (sometimes called a bolide) is a meteor brighter than magnitude 4 – i.e. a meteor brighter than Venus at its brightest. Such meteors are seen fairly frequently, and accurate and careful observation of them can yield much valuable data.

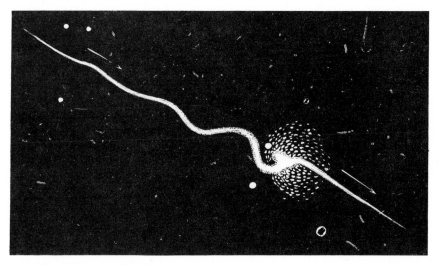

Explosion of a fireball

This is J. Silbermann's brilliant interpretation, made in 1869, of a fireball at the moment of explosion.

A meteor becomes luminous at a height of about 150 km when it begins to penetrate our atmosphere. The light fades as the speed of descent is reduced, and more often than not the meteor explodes at an altitude of about 25 km as a result of the enormous stresses built up within it.

When you see a fireball the following information should be recorded carefully and immediately after the event:—

—   Time of appearance, using Universal Time (U.T.)
—   Your position when the fireball appeared (if possible, your latitude and longitude, which can be obtained from a large-scale map)
—   The state of the sky: was it cloudy or clear? What was the limiting magnitude (the magnitude of the faintest star you could see)?
—   The path of the fireball across the sky. This can be described in several ways:
    (a) the path relative to the star background (a sketch is always helpful), giving the R.A. and Dec. of the start and end points of

the luminous trajectory, and any notable events, such as a sudden fall in brightness;

(b) the altitude and azimuth of the start and end points (in the case of a daylight fireball this is the only reliable technique); or, as the last resort,

(c) notes on the compass course on which the fireball was travelling, together with a rough estimate of its altitude above the horizon.

— The physical characteristics of the fireball:

(a) colour and any changes which occur along the trajectory;

(b) structure and shape of the head;

(c) any train forming and remaining behind the head – in particular its duration of visibility in seconds and whether or not it drifts from its original path and becomes distorted in shape. Drawings made at regular, frequent intervals are useful in evaluating structural changes;

(d) the velocity of the fireball, estimated from the apparent path, length and the duration of visibility (in seconds).

Not all fireballs produce meteorite falls, and those that do usually consist of fragile material which readily undergoes fragmentation. In order to determine the location of a possible impact site so that a proper search can be organised, it is necessary for a fireball to be observed accurately from more than one location. Calculations are then performed to determine the actual line of the trajectory in the atmosphere, which can then be 'projected' forward mathematically to meet the Earth's surface at an estimated point of impact.

Any fireball observed or meteorite fall witnessed should be reported immediately for prompt scientific investigation.

Meteorite photography and recovery projects

The chance to photograph a fragment of interplanetary debris as it enters the Earth's atmosphere is a rare experience, but just occasionally it does happen (see page 75).

Take, for example, the case of a New Mexico cowboy, Charlie Brown, eating his breakfast one early morning in March 1933, when he suddenly saw a brilliant light through the window. Grabbing his cheap, simple camera as he ran outside, he was in time to record a meteor in all its splendour, and so to make photographic history. The contribution to meteoritic science was even more valuable because the picture of the meteor in flight recorded details which could not be captured by a mere observer.

Catalogues of meteorites prosaically record this fall in the following manner:
Pasamonte, Union County, New Mexico, U.S.A.
Fell 1933, March 24, 0500 hours.   36°13′N, 103°24′W
Stone. Eucrite.
75 stones, with a total weight of 4 kg, the largest specimen 300 grams, were collected along a track of 46 kilometres.

Dr Nininger was soon on the trail. He calculated the likely flight path and instructed ranch hands to search for fallen meteorites and how to recognise them. The hunt was successful but the finders kept raising their prices until Dr Nininger was forced to pay $3.00 an ounce instead of his usual rate of a dollar a pound. The type of stony meteorite found, however, was so rare that he stayed in the area for several weeks to get as many specimens as possible.

Nowadays, when a meteor descends, trailing a shower of sparks that disappear at about 20 km above the Earth, it is likely to be successfully filmed by a network of camera stations. Time exposure photographs of fireballs now permit us to study the smoke trails left hanging in the sky and to estimate the fragility or the volatility of the meteor mass. Probably before very long automatic cameras will be coupled with sensitive microphones to record the sounds of a falling meteorite – resembling a sonic boom with successions of high pitched cracks followed by a deep rumbling.

The recording of meteor phenomena entered a new era in 1959, at which time two camera systems in operation in the Ondrejov Observatory in Czechoslovakia photographed a large fireball that subsequently dropped a shower of stones of which 19 were recovered.

The recovery was possible because the film gave enough information to calculate the meteor's orbit and impact area.

In 1964, a system of 47 camera stations was set up, covering more than a million square kilometres of West Germany and Czechoslovakia, and in the years that followed, stations built for meteorite tracking and recovery were no longer a novelty.

In 1970, the Dominion Observatory in Canada, built a network of 12 observing stations in the prairies of Alberta, Saskatchewan and Manitoba, each having five sky-watching cameras which run automatically throughout the hours of darkness. The prairies were chosen because the land is easy to travel and search, particularly in winter when lakes are frozen and snow covers the ground. Photographs from at least two stations (which are spaced 200 kilometres apart) are required to calculate the meteor's orbit and likely point of impact.

But it was not until 1977 that the vigilance of the Canadians was rewarded, when six meteorites with a total weight of 3.8 kg were recovered in the snow near Innisfree, Alberta (see page 48).

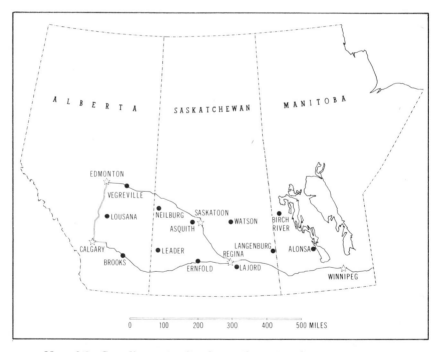

Map of the Canadian meteorite observation network

The twelve stations (filled circles) of the Canadian Meteorite Observation and Recovery Project (MORP), which became operational in 1971, are controlled from a field headquarters in the University of Saskatchewan. Its stations, which cover an area of 700,000 square kilometres, were deliberately sited away from the glare of large cities, and each one is equipped with five cameras, a meteor detector and exposure control system.

Canadian observation station

The observatory is designed to withstand winter conditions. Each of the five cameras per station has a 50mm wide-angle lens and is loaded with 100-ft rolls of film; a 'sun-shutter' in front of the lens protects it during the day-time.

The cameras are tested once a month by using a distress signal rocket flare to simulate a meteor. Each observatory is visited twice a week, but an eye-witness report of a meteor would cause an immediate visit and examination of the exposed film.

Snowmobile search

The two meteor detectors at Vegreville and Lousana in Alberta were triggered-off by a meteorite on February 6th 1977. The estimation of the impact point, from the recorded observations, was so accurate that the meteorite was discovered after only four hours of searching some twelve days later. Called 'Innisfree' after the near-by town, it is seen in front of the third snowmobile from the left of the picture. It had penetrated 30cm of snow to the frozen soil beneath, and rebounded to the surface.

This Meteorite Observation and Recovery Project is operated by the National Research Council of Canada, and covers 700,000 square kilometres – which represents one half of one per cent of the total land area of the world. The headquarters are in Saskatoon, where about 35 kilometres of black and white film are developed and examined each year.

A very sophisticated monitoring network – unfortunately no longer in operation – was also set up in the United States which very frequently recorded meteors on film, and from its findings it was deduced that, a few seconds after entry into the atmosphere, they broke up and disintegrated into very fine dust. The network waited impatiently for six long years before a meteor finally let fall a meteorite near Lost City, Oklahoma. The fall is recorded in the following manner:

> Lost City, Cherokee County, Oklahoma, U.S.A.
> 36°0.5′N, 95°9′W
> Fell 1970, January 3, 20.14 hours 17 seconds, Central Standard Time.
> Stone olivine-bronzite chondrite (H5).
> Following observations on the Prairie Photographic Network, a search area was delimited and 4 fragments totalling 17 kg recovered during the 6 days after the fall. The largest stone weighed 9.83 kg.

The Russians have also been maintaining a network, while in Britain a relatively small network to watch fireballs has been set up by amateur astronomers, who have photographed a meteor body over the Scottish Highlands, travelling at a speed of 38 km per sec (136,800 kph). The body was estimated to weigh 11 kg and supposedly ablated to a few grams in just two seconds.

Studies of films made by camera networks show that irons and stony irons have a fair chance of surviving the passage to the ground. But the fragile cometary debris is believed to be completely destroyed in the atmosphere, and only the most resistant and slowest-moving will occasionally reach the ground.

For instance, a bright meteor-fireball was recorded by the Czech network in 1974. The estimated weight was 200 tons, and the entry speed 27 km/sec (97,200 kph), but in 3.5 seconds the fireball was reduced to finest dust.

Calculations as to the number of meteorites landing on Earth vary enormously. Some estimate that millions of meteoroids enter Earth's atmosphere every year, and that only 200 reach the surface as meteorites, about ten of which are found. But more precise figures should become available as observation and recovery projects monitor the influx of meteorites, with hundreds of miles of film maintaining a continuous record.

The recovery of 'Innisfree'

Eleven years after the first planning of Canada's observation network, the filming of a fireball led to the recovery of a meteorite. In this instance, the atmospheric conditions were favourable and the cosmic material was not entirely consumed before reaching Earth. The ellipse of the fall was small (400×500 metres), and of the various fragments subsequently recovered, one was found on the roof of a cruising station wagon.

Once on the ground a meteorite absorbs moisture and is subjected to contamination, so the sooner it is recovered the better condition it is in for laboratory work. In addition, its faint radioactivity decays and alters in a short time, so measurements should be made as soon as possible. One of the great advantages of these tracking networks is that by speedily establishing the impact area a search can be mounted without delay.

What is a meteorite worth?

"Meteorites are truly far more precious than diamonds, because they carry cryptic messages of happenings somewhere in the solar system more than four billion years ago".
Fred L. Whipple, Smithsonian Astrophysical Observatory,
Cambridge, Mass.

The "stones from the sky" have been prized since earliest times – it is only the kind of value placed upon them that has now changed. Venerated in ancient days as objects of divine mystery, they were kept in the possession of rulers, high priests or medicine men. To-day, the average citizen still regards a meteorite with this sense of awe; but added to the legendary value, is now an awareness of its scientific importance, and in consequence it also has a real monetary worth in the market-place.

Natural history museums first started to collect meteoritic material in a systematic fashion some 150 years ago. By 1900 dealers were making expeditions to procure fossils, minerals and meteorites, and indications that a meteorite had fallen were eagerly followed up.

The turn of the century brought in a period of intense interest in the natural sciences, both in Europe and the United States, and natural science institutes started to publish catalogues listing meteorites for sale. One illustrated catalogue issued in 1904 listed 40 different types, in three main groups – siderites, sideroliths and aeroliths (irons, stony irons and stones). Around that time a small Canyon Diablo fragment of 40 grams was priced at $1.50 in one American catalogue; an 11 gram meteorite from Salt River, Kentucky, was listed at $10; a 505 gram meteorite from Clacksman County, Oregon, was listed at $200, and a meteorite of 142 grams named Veramin, from Teheran, was listed at $250. The largest fragments of a meteorite would find a permanent home in a museum, while the smaller stones would be sold to collectors, who, even at that time, had a great appreciation for meteorites as well as other rocks and minerals.

Since those days the development of space research and travel has given enormous impetus to the search for fresh meteoritic material. "The fallen stars must", says Dr E. L. Fireman of the Astrophysical Observatory at Cambridge, Mass., "continue to be used for research, which has already added significantly to our knowledge of cosmic rays in the outer reaches of the solar system; the development of the temperatures and pressures to which meteorites are subjected; the abundance of the various chemical elements; and of the time lapse between the formation of elements and the formation of meteorites."

The hunt is now well and truly on, and when a new fall is reported enterprising individuals can often beat representatives of institutional bodies to the site and stand a chance of getting hold of a good part of the material – which then often passes through several hands before ending up in official collections. Individual collectors competing with museums and research institutes when meteorites come on the market can usually conclude a transaction much faster than can an institute, and when Sotheby's and other major auction houses have on occasion offered meteorites for sale, they have mostly been purchased by private buyers.

'Space' jewellery is also in demand, and this will further diminish the supply of meteoritic material and increase its market value – although one cannot be dogmatic about precisely what this is, as one can about other commodities regulated by supply and demand. But the total amount of mined gold in the world, according to statistics provided by London bankers, is thought to be some 75,000 tons; while the amount of meteoritic material catalogued and in the possession of museums, State or national institutes, etc., is less than 1 per cent of the total amount of gold. In addition, the total estimated weight of all gem quality and industrial diamonds so far mined exceeds the weight of all meteorites in museum collections.

In former days when meteorites were of great interest to only a few, they could frequently be purchased in areas where large amounts of such material were easily available. Museums, generally, were content to secure a few of the larger fragments, and were not interested in searching for additional specimens. Finders could not expect large rewards, and they were usually farmers, unable to travel to places where they could most profitably market the meteorites, so they just sold them to anyone who made a reasonable offer.

Now that this cosmic material is becoming increasingly desirable and valuable, the main factors to be taken into consideration are:—
— The quantity recovered – how much is left for sale to the public after the larger fragments have been secured by museums?
— Certain meteorites have a rare structure and composition and are highly valued (in some cases there may be only two or three known specimens).
— The number of separate specimens in the fall or find will affect the price – sometimes several hundred stones are recovered, sometimes only one.
— For certain research purposes the value of the meteorite is greatest immediately after its fall, before it is subjected to contamination and oxidisation.
— Precise data on the fall increases its value – e.g. does it have archeological associations, has it been found in a burial site, etc?
— The surface characteristics of the meteorite may be important.
— Specimens of exceptional beauty are more highly valued.
— The geographical location of a 'fall' may be important.

Not long ago the Canadian Government, through the Astrophysics Branch of the National Research Council, offered to buy any new meteorite found in Canada – the price to depend on type, size, fresh-ness, condition and 'fall' data. So the value of a meteorite is therefore largely arbitrary. One particular stone offered on the market may be the last for a long time, with replacement virtually impossible. Museums and research institutes are sometimes forced to exchange samples of no current interest for those needed for a specific research project, and this is how some collectors make fresh acquisitions: by agreeing to make an exchange in order to obtain a meteorite of which they have no specimen.

Collectors who possess all the available specimens of a particular meteorite are in an especially strong position vis-à-vis museums and research institutes. The author has spent much of his life studying and collecting meteorites: some of those he sold have subsequently increased immensely in value – for example, the important Murchison meteorite from Australia (which, as mentioned in another chapter, was found to contain amino acids). Meteorites that are of little interest to-day may suddenly spring into the limelight through the discovery of a clue that opens the door to a whole new world of scientific knowledge. Each meteorite is therefore a potential treasure house of knowledge, a factor that must always be taken into account in any assessment of its value. Moreover, although official tracking networks now exist in many countries, the rate of recovery of fresh material is still only a few meteorites a year.

At one time there was little general interest in these stones – dealers could not sell them because they were not 'beautiful' enough for the average collector of minerals. Meteorite collection was then the sole preserve of those who were 'in communion' with these mysterious objects from outer space and sought to divine what messages they might contain. The meteoritic material which then came on to the market from 'productive' areas, at fixed prices per gram, is just not available now. Good bargains were only to be had while the possessor did not realise the value of what he had, and a farmer would usually accept a modest offer for meteorites found in his fields because they meant little or nothing to him. In fact, at the turn of the century some iron meteorites ended up in foundries where they were smelted for their metal content.

But even though prospectors and collectors have become increasingly aware of the value of their finds, and thus increasingly reluctant to sell, some of their collections have been sold to museums or scientific institutes for only a fraction of what they would have fetched on the open market. The individual owner of a meteorite is confronted with various possibilities: a museum might make a reasonable offer, but if the meteorite is particularly important for research purposes, the owner can often get a higher price for it elsewhere. If the owner has detailed documents and keeps abreast of the literature, he will certainly be able to make a much better sale than someone who may know only that the stone he has is a meteorite.

Are finders, keepers?

"Catch a falling star and put it in your pocket ..."

But follow the advice of this popular song and you may find yourself in trouble.

Most land belongs to somebody, either a Government or a private owner, and even that which appears to be 'public' property is likely to be grabbed by the State if it suddenly proves to be of value – even if retrospective laws have to be passed to enable them to do so. So, after a meteorite has been found and identified, the question is – to whom does it belong? There is, of course, no easy answer, as the pertinent laws (if any exist) vary from country to country.

In the United Kingdom, Parliament approved a bill in 1971 proclaiming all meteorites falling on national territory to be the property of the Crown. In Australia some States have passed similar legislation, and prospecting around meteorite craters on government land and in national parks is forbidden.

With or without legislation, finders of newly fallen and very large meteorites can hardly ever expect to keep them. But very much will depend on whether these land on public or private property. Where there is no existing legislation, the case would probably be decided on the basis of the property laws currently in force.

In 1890, in the U.S.A., Peter Hoagland received permission to search for a meteorite which, when found, weighed 30 kg. Unfortunately the permission had been granted by the tenant of the land, and not by the owner, who then claimed the meteorite, took the case to court, and won.

Another example is the famous case of the Willamette meteorite, in Oregon, U.S.A., which was discovered in 1902 by Ivan Hughes, a local farmer, in a dense forest there. His solid iron-nickel treasure weighed over 13 tons, and his problem was how to move it secretly onto his own land. He constructed an amazingly ingenious contraption, using great ropes and pulley wheels cut out of the tree-trunks; working painfully and slowly with his horses, under cover of darkness, he finally got the meteorite onto his own land – where he thought it would be safe. But the forest belonged to the Oregon Steel Company. They took the case to the State Supreme Court, which ruled that meteorites go with the ownership of the land. The meteorite was subsequently sold for $20,500 and is now on display in the American Museum of Natural History in New York. Poor Ivan Hughes, despite all his efforts, got nothing.

In 1930 an iron-nickel meteorite weighing between 25 and 27 tons was found in Tanganyika. It was promptly declared to be an 'ancient monument', and no piece could be removed without the special consent of the Monuments Commission. Similar restrictions cover other 'giants', such as the iron-nickel Gibeon meteorite shower in Great Namaqueland, South West Africa. The largest known meteorite in the world – found near Hoba West in Namibia (South West Africa) and weighing some 60 tons – was proclaimed a national monument by the South African Government just in time to save it from the smelters.

Modern prospecting can be a costly business, involving light aircraft, cross-country vehicles and sophisticated equipment – to say nothing of a great deal of the hunters' time. So before making such an outlay it is only prudent to make sure in advance to whom any finds would belong, and this can be further complicated by the fact that a good area might attract several competing teams.

One practical arrangement is for a prospector to pay the land-owner a small fee in advance for the right to keep any meteorites he may find, or to agree that the selling price be divided beween the finder and the land-owner.

But in spite of all the laws and precedents, most of us cannot help feeling that if we find something that has fallen out of the sky it ought to be a case of finders – keepers!

Precious meteorites

Twinkle, twinkle, little star, ...
Like a diamond in the sky!   Jane Taylor

Amid a blaze of light and loud explosions, several meteors fell a few miles from the village of Novo Urei, on the banks of the Alatyr in Krasnoslobodosk district of Russia on the morning of September 10th, 1886. Some local peasants were the first to find the meteorites, and breaking one of them open, they solemnly proceeded to eat it. There is no record of what this did to their digestion, or to their teeth; but subsequent examination by the scientists Yerofeyev and Lachinov revealed that the meteorites contained diamonds! They classified the stones as olivine-pigeonite achondrite, and coined the name 'Ureilite (Novo Urei)' for them.

The Russians, whose agate mortar was deeply scratched when they tried to pulverise a small fragment for analysis, found scattered carbonaceous matter in the meteorite, and investigation revealed that the small greyish grains were, in fact, carbonado (black diamond, minute crystals of which are used for industrial purposes). The diamonds were present in the amount of 1 per cent, and since the meteorite weighed 1762.3 grams, the diamond content was 17.62 grams, or 85.34 carats. There were a few minute clear specimens among them.

Since this first discovery there have been other instances of minute diamond crystals in meteorites – e.g. the Magura iron meteorite in Czechoslovakia, and the Carcote meteorite in Atacama, Chile.

Recent investigations have shown that many meteoritic iron ores contain small cubic crystals of graphite, now christened 'cliftonite'. Scientists have speculated that these crystals might originally have been diamonds which were converted into graphite by heat. Diamonds found in meteorites have been yellow, black, blue, or even colourless, but always in minute or microscopic amounts. Many of the meteorites lodged in museums have only been studied in a cursory manner, so it is quite possible that further investigation will reveal the presence in them of diamonds.

The most famous meteorite crater in the world, Canyon Diablo in Arizona, has yielded carbonado in copious amounts, with the result that people seeking to cut Canyon Diablo meteorites only succeed in destroying their grinding wheels. The presence of diamonds in Canyon Diablo meteorites is unpredictable, but it is believed that there are some 12 minute diamonds per kilogram of material, and polishing several hundred sections of these meteorites revealed one caronado inclusion per 206 square centimetres of polished surface, the

Diamondiferous meteorite

This fine specimen of a siderite in the author's collection bears the unmistakable 'signature' of a Canyon Diablo, Arizona, meteorite. Its dimensions are 44 × 32 × 22cms and its weight of 102 kilos makes it among the twenty largest specimens of this fall.

Reckoned to have fallen a mere 20,000 years ago, its cosmic age is estimated to be 1.2 billion years. Its predominant constituent is iron, with nickel and traces of cobalt, copper, chrome, platinum, gallium, germanium, palladium, and gold. It also contains minute diamonds embedded in pockets of carbon.

diamonds ranging from 0.05 to 2 millimetres in diameter. These carbonados were embedded in graphite, and the crystals are transparent, yellow, brownish or white (see page 73).

It was Dr A. E. Foote, a minerals dealer from Philadelphia, who first reported the presence of diamonds in the Canyon Diablo meteorites in 1891. These diamonds have no commercial value, but their presence in the meteorites quickly captured the imagination of the general public, and the Canyon Diablo became the site of a veritable 'space-diamond' rush. Another attraction in the Canyon Diablo meteorites were the minute quantities of cobalt, platinum and other elements. They also contained two rare minerals not found in terrestrial rocks; Cohenite and Schreiberite, which reputably make the meteorite fragments in which they are found difficult to cut.

The diamonds in the Canyon Diablo iron meteorites were confirmed by X-ray tests a long time ago, but their properties and formation processes are still being studied for the valuable information that they contain.

Despite their microscopic size, meteorite diamonds are highly prized by collectors because of their origin in space. Dr Samuel Tolansky, a British scientist and the author of many books on diamonds, believes that pressures and temperatures similar to those needed to manufacture synthetic micro-diamonds could have been produced by the impact of giant meteors on the Moon. Since the Moon has no atmosphere, the falling meteors would not be retarded, and the graphite would be turned into diamond; indeed, the diamonds thus produced would be larger than those created by the impact of meteors on Earth, or in laboratories. This has not so far been confirmed by the samples collected by our astronauts from a very small area, but Dr Tolansky suggests that there could, in fact, be acres of diamonds on the Moon.

'Gems of the Gods'

In early times meteorites were for the most part regarded with too much awe to be used merely for personal adornment. They were, rather, objects of veneration – like the Abdhuta-Nāth, in Bengal, where a Brahmin temple was specially built to house this meteorite. Gradually, the 'gems of the Gods' were also converted into weapons which, by virtue of the divine properties attributed to the metal from which they were made, conferred on their owners supernatural powers and invincibility. The first conquerors of Europe – Attila the Hun among them – were reputed to wield 'swords from heaven'. In the 19th century, Alexander of Russia was presented with a sword forged from a section of the Cape of Good Hope meteorite found in 1793; and the Japanese court possessed several swords made from the Shirihagi meteorite dredged up in 1890 from the bed of the Kamiichi-kawa River.

But with the passage of time, more and more interest has come to be shown in fashioning personal jewellery from meteorites – jewellery prized not only for its far, mysterious origins, but also for the rarity of its material, and for the beauty with which this can be wrought into unique pendants, rings and necklaces. Meteorites usually have an uncouth appearance in their raw state, and all the talents of the designers and craftsmen are needed to transform the material into glamorous jewellery. Nevertheless, the striations on certain metallic meteorites are beautiful in themselves, and the Widmanstätten pattern which emerges with mild heating and hammering is lovely enough to inspire the most demanding designer, as is the result of etching the polished surface.

There is also, of course, the intrinsic beauty of the materials contained in some meteorites – gold, platinum, copper, diamonds, and – above all, perhaps – peridot, the gem variety of olivine, which, as far as we know, is the only true gemstone of celestial origin. Certain meteorites contain grains of yellowish-green olivine, and some have green and transparent peridot – though the crystals of the latter are very small, and there are only a few meteorites with sufficient peridot for the cutting of a one-carat stone. The author suggests the name 'cosmite' for this truly celestial gem which, unlike terrestrial peridot, contains no nickel. 'Cosmite' is, of course, unobtainable on the open market, and private collections and museums preserve what little there is. Peridot occurs in the metallic network of pallasite meteorites and, with only 2 per cent of all meteorites being in this group, they are much too rare to be used for jewellery.

In fact, with most meteorites in museums or private collections, and with so little material at hand for making jewellery, we shall have to wait until space ships bring back cargoes of minerals from asteroid mines for any production of ornaments on an industrial scale.

The different characteristics of other meteorites have already been valuable, of course, in other ways. Iron meteorites containing nickel and other elements in certain proportions have extraordinary strength, resistance to corrosion, and durability. The manufacturers of the armour-plating for battleships imitated the composition and structure of meteoritic iron, so as to obtain the maximum resistance to penetration, the material being given the name 'meteor steel'. For more general use, stainless steel is, to a great extent, patterned upon a meteoritic alloy; while the irridescent effect achieved on some watch dials and other surfaces was directly derived from metallic research on the structure of meteorites.

There has recently been a development of combined educational exhibits under the patronage of commercial, cultural and scientific interests. This collaboration has permitted meteorites to be displayed in a way which makes it possible for a wide audience to appreciate their many values – historical, scientific, practical and artistic.

In the forefront of this collaborative movement are Schlegel & Plana, of Berne, Switzerland, who sell meteorites set in magnificent symbolic mountings (see page 79). Sometimes their designers recreate the flaming descent of the meteor to earth, using gold to depict the glow that lit the meteor to incandescence as it entered our atmosphere. They also use iron-nickel meteorites in their free form.

Schlegel & Plana are careful to provide their meteorite products with a certification of cosmic origin, together with very complete information about their archaeological data, composition, cosmic age exposure, terrestrial age, etc. Although most meteorites used for space jewellery have already been thoroughly studied, it is always possible that some one particular meteorite could, at a later stage, contribute further insights into the unravelling of the mysteries of the solar system. Thus it is of the utmost importance that each owner should have and preserve this documentation, for future researchers into qualities and modes at present not even guessed at.

There is a current travelling exhibit, 'Man and Meteorites', in which the 'pole d'attraction' is a display of the most modern techniques for tracking meteor orbits and predicting impact areas. This sophisticated technique cannot pinpoint the precise fall, so we count on the zeal of hundreds of local school-children, scrambling over the terrain, eager to be star-finders, to locate the scattered meteorite fragments. These children do not by any means constitute an unskilled labour force: they are given sufficient briefing to distinguish a meteorite from an earth rock, and an experience like this is something to remember and to provide an interest for the rest of their lives.

Meteorites in art and literature

Since man first began to recreate the marvels of the world around him – using chalk, and the cave-wall for canvas – fiery-tailed meteors and shooting stars have figured in his art. The earliest depiction of meteors so far discovered is in the pre-historic cave paintings in Altimira. In later times, the temples built to enshrine the "iron that fell from heaven" were often decorated with illustrations of its marvellous descent.

Paintings have become a part of the documentation of past meteorite falls (see pages 33 & 34), even though artistic licence has represented them as fiery chariots, dragons, fire-horses and other imaginative wonders; in fact, careful astronomical observation of the heavens has had little to do with mythology and art, Tintoretto, for example, created a cosmic myth in his 'Origin of the Milky Way' in which stars burst forth from the breast of a goddess suckling an infant god!

Perhaps the most famous representation is Raphael's 'Madonna di Foligno', in the Vatican Art Gallery in Rome (see page 37). Behind the Madonna and Child, Raphael has painted low in the sky a brilliant fireball with a smokey red tail. The richly-coloured painting is one of the glories of the Vatican Gallery, and it is also a source of speculation to art historians. The most generally accepted story is that a Papal Secretary commissioned the painting to celebrate his escape from a falling meteor. This was probably the one that fell at Crema in 1511 – one stone of which was reputed to weigh over 50 kilos. A contemporary historian has left this account: "The illumination was bright enough for the people of Bergamo to see the whole plain of Crema during darkness".

Not only chroniclers but poets of all ages have been fascinated by falling stars and meteors, and they have also inevitably found their way into folklore and legend. The Lithuanians, for example, have a myth which relates them symbolically to life and death. "When a child comes into the world, Wepeja spins the thread of his destiny, and each thread is attached to a star. At the instant of death the thread is broken, and the star falls, pales and fades away".

Homer and Vergil were the earliest of the poets of all civilisations to sing of shooting stars. For some since, it is the sense of their beauty that has been uppermost – most notably, perhaps, for Shakespeare, for the birth of whose lovely heroine Beatrix "there was a star danced"; or for Shelley, intoxicated with "light, in star-showers thrown". Others, John Donne and Browning amongst them, sought in the stars some metaphysical significance. While yet others, Thomas

Middleton for one, read there a personal destiny – "Beneath the stars, upon yon meteor, Ever hung my fate, 'mongst things corruptible".

Pope Pius IX marvels at a meteorite shower falling over Rome in November, 1872

Contemporary poets no less than earlier ones have been fascinated by the mystery and beauty of falling stars. And – after a lifetime devoted to them – it is hardly to be wondered at that they have even inspired a scientist to poetry.

Dr Nininger – whose 'A Meteorite Speaks' thus:—

A hieroglyphic message is written on my face
Recording ancient happenings far in the depths of space.
It tells of my beginnings where fiercest fires held sway,
My leap into ethereal space and how I sped away.

A diary of my wanderings, lonely 'mongst the stars,
A thousand of such incidents as Jupiter and Mars.
I've watched a host of planets grow from out the spatial voids;
Witnessed lunar peltings and played tag with asteroids.

I held my course through solar heat, likewise through frigid
    space,
I wooed the lovely Pleiades and gave Orion chase.
I knew severest loneliness from all celestial forms;
Likewise the social gaiety of cometary swarms.

Freely through ethereal space I loved my course to steer,
But trapped at last, fell victim to earth's dread atmosphere.
In arid wastes I landed, then, smote by desert sand
My skin deep brown was varnished by oxygenic hand.

                                          Dr H. H. Nininger

Fact is often stranger than fiction

The representations of meteorites and comets by early artists and writers were, as often as not, weird imaginings worthy of Dante's Inferno, with hordes of demons and monsters, but lacking any scientific basis: pure fantasy, in fact. But all this changed with the advent of Jules Verne, the pioneer of science fiction, who supported his tales with a wide variety of disciplines, including geology, astronomy and physics. In 'Off on a Comet', written as early as 1877, he provided wonderful scenic descriptions of Jupiter, Saturn and numerous asteroids – and some of his fantasies have proved to have been uncannily near the truth.

'The Hunt for the Meteor' is Verne's tale of a scientist who invents a 'neutral helicoidal ray' in order to attract to Earth a solid gold meteor which had been observed in the heavens. It lands in Greenland, thus making potential millionaires of all its inhabitants; and potential nonsense of the world's banks and stock exchanges. But, appalled by the greed of all those who are determined to profit from the golden meteor, the scientist alters his calculations so that it is deflected into the sea and lost forever. By a strange coincidence, in the year 1908 in which the book was published, a real meteor crashed with terrible force in the forests of Siberia – an event which is described in another chapter of this book.

By the time Neil Armstrong set foot on the moon in 1969, many science fiction novels had been written in which meteorites played a significant part. Space art, too, depicted these celestial wanderers, and the results were sometimes so weird that they appealed only to a very limited public. But, as time went on, fact was increasingly inter-twined with fiction, and scientists themselves derived stimulus from the imaginings of science-fiction writers and artists.

The general public, too, is becoming increasingly appreciative of space fiction and art now that what was once thought to be wild fantasy is known to be within the bounds of reality. It will accept without question a script or scenario in which the Earth becomes the target of a never-ending meteoric bombardment, just like other planets; or where teams of astronauts are sent out to 'capture' huge asteroids and bring them back to Earth. Nowadays fact *is* often stranger than fiction.

A recent example of the space film – 'The Meteor' – clearly shows the degree of sophistication and scientific verisimilitude which this genre has now reached. It describes the threat of Earth's annihilation

Illustrations from Jules Verne's 'The Hunt for the Meteor'

by collision with a giant meteoroid, and succeeds in adding to the sheer suspense of its theme, the 'moral tone' of a collaboration between the super-powers which, alone, averts calamity.

The film contains a surprising amount of 'science fact'. For example, the asteroid Icarus is known to measure over a mile in diameter, and if it were to leave its orbit and hit the Earth much of our population would perish. A previous chapter has already described the devastation caused by the impact of a small comet in Central Siberia in 1908. There was another threat of major dimensions on February 12th 1947, north of Vladivostock. Providentially the huge cosmic body exploded in the atmosphere, so only the fragments fell on Earth in the form of iron rain.

Space phenomena will continue to provide artists with rich material for spectacular films and novels; while the products of their imaginations will enhance the work of the scientists who turn their visions into reality.

Apocalyptic vision

The destruction of the world after collision with a comet as depicted in Heinrich Harder's 19th century fantasy.

What happened to the dinosaurs?

65 million years ago the dinosaur was the largest and most terrifying lord of the land animals. Then, in a 'moment' of geological time it disappeared from the face of the Earth. Since the discovery of this fact many and varied have been the explanations offered.

One theory holds that a vast star explosion occurred near enough to Earth to wipe out many species by cosmic rays. Another, that the great, slow-moving dinosaurs were surpassed by newly evolving, faster and brainier species, to which they fell victim. Such obvious causes as disease have also, of course, been considered.

But exciting new evidence gives us reason to believe that at about the time of the dinosaurs' demise an asteroid about 10 kilometres in diameter crashed into the Earth, creating a vast crater some 150 kilometres in diameter, and blasting clouds of dust particles into the upper atmosphere. The effect of this would have been to form an impenetrable layer blocking out the sun's rays and stopping the pattern of photo-synthesis for a number of years. As food grew scarce, the dinosaurs with their voracious appetites would have been the most vulnerable species and the first to die out, whereas the smaller animals could survive by foraging for roots and seeds. In the years that followed, the dust settled and many dormant plants gradually revived, but the dinosaurs were gone for ever.

In support of this theory, we have only to remember the nuclear bomb tests of the late 1950's, when large amounts of radio-active particles were injected into the upper atmosphere, where they remained covering the entire globe for well over a year.

During the Krakatoa volcanic explosion of August 27, 1883, an entire island between Java and Sumatra disappeared, killing 36,000 people. A cloud of ash spewed 80 kilometres into the stratosphere and the blast was heard 5,000 kilometres away. It has been estimated that 18 cubic kilometres of material was blown out of the volcano, and that about 4 cubic kilometres remained in suspension in the stratosphere for a long period of time. This massive pollution created fabulous sunsets, not just locally but throughout the entire world, and light intensity was reduced and temperatures dropped.

More recently, the 1980 eruption of Mount St. Helens in the United States created a belt of fine particles about 2 kilometres deep at a height of 18 kilometres above sea-level, which is about 30 times higher than the average particle layer. Mr Hirona, a geophysicist at the Kyoushou University of Japan, who carried out these measurements using radar and laser techniques, is now tracing the subsequent weather patterns, which are revealing continuing climatic perturbations.

Such incidents do, at least, illustrate the far-reaching effects of pollution in the upper atmosphere, and lend credence to this as a possible cause for the disappearance of the dinosaurs. But even more powerful and directly related scientific evidence has recently come to light.

In the 1930's a Swiss paleontologist, Otto Rence, identified and studied a layer of clay that had been deposited at the time of the death of the dinosaurs. Subsequently, Luis W. Alvares, a Nobel Prize winner, and his son Walter, a geologist, working with a small team of colleagues, found that the fossil record at this particular stratum in the clay showed a sudden increase in iridium. Analysis has revealed that meteorites contain many times more iridium than most earth samples, and that it is found in abundance in carbonaceous chondrite meteorites. In fact, by measuring the amount of iridium deposited in this critical layer throughout the world, we can estimate the size of the asteroid needed to produce it.

This, in fact, is the keystone of the new theory, and so far no solid evidence to contradict it has been produced. However, the author points out that there is another possible explanation for this distribution of iridium in the critical stratum. At the relevant time, the Earth passed through an exceptionally dense and fast-moving swarm of meteroids. We know that the faster meteroids become luminous at the highest levels above the Earth and that the point at which their light disappears is also very high. This means that the body of the meteroid is burnt out and the fragments shower down over a vast area. Although this could account for the more or less even distribution of iridium in this particular clay layer, it offers no solution to the mystery of the dinosaurs' disappearance.

All in all, the asteroid impact hypothesis is still the most attractive, but more research, involving the accurate dating of fossils and sedimentary rocks, and the newest scientific techniques, remains to be done before the theory can be wholeheartedly accepted. For example, recent developments in satellite photography have enabled us to identify impact structures which were not visible to us before, and three craters over one hundred kilometres in diameter have already been found. These are not the correct age to have played a role in the disappearance of the dinosaurs, but even if we cannot locate the crater that meets the required criteria for this, it may only be because it is now covered by an ocean. Moreover, the new technique called paleo-magnetism – which shows how the Earth's magnetic field was orientated at the time when the sediments were deposited – could well prove to be of prime importance.

Mining in space

'The High Frontier Feasibility Act' was recently passed by the U.S. Congress. The title describes – in a strange mixture of pure poetry and bureaucratic caution – what may be the foundation stone of another Industrial Revolution, of cosmic proportions. The Act was designed to give support to a study of space colonization and mining possibilities, for which entirely new technologies would have to be developed. At the moment not a high priority is being placed on this programme, but a break-through by a competing country would certainly provide the stimulus for renewed activity at high official levels.

Samples of Moon soil brought back to earth contain, among other things, 20 per cent silicon, 12 per cent aluminium, 4 per cent iron, 3 per cent magnesium, and up to 5 per cent titanium. Because of the Moon's low gravity, no deep-shaft drilling would be required, and the moving of large masses of ore would not be difficult. Sending anything out of our gravity field requires enormous energy – for instance, the modest French Ariane project uses one ton of fuel per second for lifting-off with a very small pay-load; but once equipment was placed on the Moon, its energy consumption would be negligible, and the Moon rocks contain enough oxygen for the life-support of working crews and their operational needs. Ore could be catapulted from the Moon into space, brought into the Earth's orbit, and then given a controlled landing.

The Apollo Moon landing cost $50 billion, but it was money well spent for the mass of information and samples brought back, and for the associated technologies which it developed. The initiation of a space mining project has been estimated to cost $100 billion; this is less than the military budget of the United States for the year 1980 alone, and the rewards from such an investment would be incalculable.

Dr David Criswell, of the Lunar Science Institute, Houston, Texas, is the outstanding authority on possible Moon mining. "The Moon's surface is so rich that a million tons of ore could be excavated annually with one automated bulldozer. Excavation depth would be very shallow – like a gravel pit." The transport unit would use magnetic impulses derived from electrical energy to propel huge buckets of ore from the Moon to a collection station in orbit. Such a unit has already been designed by Professor O'Neill of the Physics Department of Princeton University, where a working model has been built. Tests conducted in a tunnel simulating space conditions showed that the buckets accelerated from 0 to 130 km/hr in one tenth of a second.

Tens of thousands of relatively small bodies, called asteroids, are in orbit between Mars and Jupiter. Larger bodies, averaging between 10 and 20 million tons, are called planetoids, and may be less than 100 metres in diameter, depending on whether the composition is basically metallic or stony. From our study of meteorites, which are fragments of these bodies, we have been able to study their content – of iron, nickel, cobalt and trace amounts of precious metals. Analyses of cross-sections of meteorites indicate that over 12 per cent of each asteroid is composed of other metals such as aluminium, magnesium, manganese and titanium.

It is not beyond the reach of our imagination and knowledge to devise programmes in which these bodies could be captured and towed towards Earth by space-ships (see page 74). Scientists and technicians are at their drawing boards working on such projects. Others 'dream dreams and see visions' of even further frontiers. The space-craft 'Voyager' has approached Jupiter and sent back precious information that has already prompted these 'dreamers' to design scoops to be used by an orbiting vehicle which could mine Jupiter's chemical atmosphere.

It is no longer inconceivable that we should make free of the mineral wealth of outer space; given the rate at which the Earth's resources are being consumed, it is only inconceivable that we should not. Indeed, great fortunes are likely to be made in the asteroid belt in the beginning of the 21st century, when space-based activities will become a routine much as off-shore drilling for oil is today.

Life in outer space?

"Is it probable for Europe to be inhabited and not the other parts of the world? Can one island have inhabitants and numerous other islands have none? Is it conceivable for one apple tree in the infinite orchard of the universe to bear fruit, while innumerable other trees have nothing but foliage?"
        'Dreams of Earth and Sky', Konstantin Tsiolokovsky

Tsiolokovsky's conviction that there is life other than ours in the 'infinite orchard of the universe' is one increasingly shared to-day by many scientists. Meteorites can provide them with vitally significant clues to the mystery of extraterrestrial life, and the importance of this research increases as more sensitive analytical instruments are constantly being developed.

The first meteorite to give rise to an extensive literature on its carbon compounds and alleged fossil inclusions was the carbonaceous chondrite which fell in May 1864 at Orgueil, in France. The catalogue entry is as follows:

        Orgueil : Montauban, Tarn-et-Garonne, France.
        Fell   : 1864, May 14, 20.00 hrs. 43°53′N, 1°23′E.
        Synonym: Montauban
        Stone   : Carbonaceous chondrite

Dr Nagy and Dr Claus subsequently reported finding in the Orgueil meteorite "organised elements which do not resemble any mineral form", and reached the guarded conclusion that the 'organised elements' might have come to Earth in the body of the meteorite. It is also known that the rare carbonaceous chondrite meteorites contain carbon compounds which are present in living organisms. Needless to say, the report sparked off considerable scientific controversy.

The physicist, Edward Fireman, of the Smithsonian Astrophysical Observatory, contended, on the other hand, that earthly life can enter the crevices of a meteorite and that carbonaceous chondrites are porous, and easily absorb moisture and organisms from the sweat of people handling them.

It is known that meteorites on display in museums over a long period of time are invaded by earthly microbes, and the possibility of terrestrial contamination has always to be taken into account in meteorite analysis. For this reason the 'freshness' of the meteorite is very important as regards reliability of deductions – especially in this matter of extra-terrestrial life.

The Orgueil controversy triggered some interesting experiments, but substantial progress was only made when a meteorite fell in

Murchison, Victoria, Australia, in September 1969 (see page 76). When the carbonaceous chondrite, type II, was examined, amino acids were found, and these are the basic building blocks of life. They represented 15 parts per million in the sections which were taken out from the inner part of the meteorite and examined with great caution to avoid contamination. Amino acids consist of chains of hydrogen, oxygen, carbon and nitrogen atoms, and different atomic arrangements result in different acids. Earth amino acids always spiral to the left, whereas in the Murchison meteorite there were right-turning amino acids. (Technically these patterns are called 'dextral' for the right and 'levoratory' for the left.) This strongly suggested that they had come from outer space, and that they were a significant pointer to the existence of extraterrestrial life.

Dr C. Ponnamperuma, a geochemist at the University of Maryland, was responsible for establishing the presence of extraterrestrial amino acids and hydrocarbons in the Murchison meteorite. He concluded that organic molecules found in it had been formed before the meteorite reached the Earth. The specimens he selected for study were ones which had a generally massive character, fewest cracks, and the least exterior contamination.

It is now accepted that amino acids are indigenous to certain meteorites and, therefore, extraterrestrial. The analysis of carbonaceous chondrite stones reveals the presence of up to 5 per cent carbon and 13 per cent water. But new insights to support the idea of 'molecular fossils' were provided by the examination of two meteorites recently found in the Antarctic, preserved and frozen in ice there for some 200,000 years. Analyses carried out by Dr Ponnamperuma and his collaborators showed the same organic composition on the outside and inside of the specimens, and their studies led to a new estimate that life existed on Earth 3.8 billion years ago, when Earth was a mere 800 million years old.

The significance of finding 'life forms' in meteorites is obvious. When NASA scientists reported, after examining the Murchison meteorite, that their finds were 'probably the first conclusive proof of extraterrestrial chemical evolution', it was the beginning of an accelerated search for evidence of life in space. Radio astronomy observations made by spectographic techniques discovered ammonia molecules, which are called 'chemical ancestors' of life, and research is now being carried on to find new molecules in outer space.

We now accept that the 'seeds of life' exist throughout the universe. We can speculate endlessly – about super-civilisations that might have visited this planet during the ice ages, during the reign of the dinosaurs, or when trilobites inhabited the seas. It is science fiction, and perhaps more fiction than science. But radio-astronomers are already beaming out signals far into the once unimagined regions of space, from which our ultimate enlightenment may come.

The Basic Vocabulary of Meteorites

A

Ablation is the wearing away by melting of a meteorite as it passes through the atmosphere. During entry into the atmosphere, the meteorite is slowed down and may break into fragments.
During its flight through the atmosphere, it is shaped by the air and develops characteristic flow marks.
Final velocity of smaller meteorites can be as low as 300 km/h.
Achondrite A stony meteorite without chondrules.
Aerolite A meteorite containing more stone than iron.
Annual Meteor Shower A meteor shower which occurs at approximately the same time each year when the Earth's orbit passes through a stream of cometary debris also in orbit around the Sun.
Asteroid is a small body found in the belt between Mars and Jupiter. Asteroids range in size from 15 m to 700 km. They are sometimes called planetoids because of their small planet-like bodies. So far over 40,000 asteroids have been discovered. The asteroid Icarus is 1.6 km in diameter. The largest is Ceres, 700 km in diameter. Eros — found in 1898 — is shaped like a sausage and measures 30 km long by 15 km.
Astroblem is an impact scar made by a large meteorite, asteroid or comet.
Astronomical Unit (AU) A means of measuring distances within the solar system. One unit is equal to the mean distance of the Earth from the Sun; approximately 93 million miles.
Astrophysics A branch of astronomy that is concerned with the physical processes at work in the universe, especially in the interiors of stars.
Ataxite An iron meteorite consisting of either pure kamacite, an irregular mixture of kamacite and taenite, or pure taenite.

B

Big Bang The cataclysmic explosion that many astronomers believe marked the birth of the universe.
Black Hole A massive star that has collapsed after running out of nuclear fuel. Its enormous gravity allows nothing to escape — not even light; hence it is invisible.
Bolide A brilliant shooting star, a large meteor, especially one that explodes.

C

Chondrite Stony meteorites in which chondrules are found. Chondrules are rounded grains composed of one or more minerals, often having a radiating structure.
Coesite A polymorph of silica produced by high pressure. Coesite-bearing sandstone fragments are found in the vicinity of large meteorite craters. They indicate that the crater is of meteoritic origin.
Comet From the Greek kometes, long-haired. A celestial body drawn within the Sun's gravitational field and occasionally close enough to the Earth to be observed. As the Earth crosses different comet orbits, spectacular meteors are observed. Meteorites are broken-up comets.
Comets are a collection of frozen gases (methane, ammonia) and solid particles that travel around the Sun in highly elongated orbits. They are usually invisible to the naked eye until they near the Sun, at which point their components are heated to produce a coma of gases and dust that becomes a tail extending, in some cases, millions of miles.
Constellation A group of stars that, in the eyes of the ancients, formed recognizable images.
Cosmic Rays Energetic particles consisting mostly of protons.
Cosmology The study of the origin and development of the universe.

E

End-Point The point where a fireball disappears, often in a shower of 'sparks'.
Etch To corrode a prepared surface with acid for the purpose of revealing structural details.
Exobiology The study of extraterrestrial life.
Extraterrestrial Beyond the Earth.

F

Falls A 'fall' is a meteorite that has actually been seen to fall and has been recovered afterwards at the place of impact.
Finds A 'find' is a meteorite, not observed to fall, but found and identified as such.
Fireball A popular term for a large

meteor, sometimes emitting meteorites.

Fission The splitting of the atomic nucleus; the process at work in nuclear reactors or atomic bombs.

Fusion The combination of atomic nuclei to form heavier nuclei; the process that powers stars and the hydrogen bomb.

Fusion Crust The paper-thin, burned layer on the surface of a meteorite, caused by frictional heat during its descent through the Earth's atmosphere. This outer covering is produced by the solidification of melted surface materials.

G

Galaxy A huge island of stars in space bound together by their gravitational attraction.

Gravitation The natural force of attraction that exists between all particles of matter throughout the universe.

H

Hexahedrite Iron meteorite consisting of large cubic crystals of kamacite.

I

Impactite Impactite is composed of meteoritic material fused with melted rock. It results when a large meteorite strikes the Earth with sufficient force to form a crater. Fragments can be smaller than a grain of rice. Microscopic examination reveals meteoritic metal.

Isotopes One of the various forms of an element. Isotopes of the same element have the same number of protons in their nuclei but a different number of neutrons. Cosmic age exposure is measured by isotope techniques. Isotopic composition of meteoritic elements has been investigated and various measurements can give the cosmic age.

Radioactive tracer techniques are said to be the most powerful analytical tool since the invention of the microscope.

L

Light-Year The distance that light, travelling at a speed of about 186,000 miles per second, traverses in a year: roughly 6 trillion miles.

M

Metallic Spheroids Tiny metallic droplets resulting from the explosion of large crater-forming meteorites. They condense from the vapour that could have been formed by the meteoritic explosion.

Meteorites Meteorites are fragments that survive the passage through the atmosphere and reach the Earth's surface. A very small fraction of extraterrestrial bodies is captured by our gravity and falls to Earth. Most of them are burned up and less than ten meteorites are recovered yearly. A meteorite shower from a single body is treated as a unit, and the fragments are all known by the same name and share the same basic characteristics. The term is also used to refer to the object in space before colliding with the Earth. Meteorites range in size and weight from tiny stones weighing mere grams to gigantic chunks weighing 60 tons or more. Meteorites have been found to be an estimated 4.5 billion years old — or about the same age as the solar system.

Meteorite Path Trajectory of a meteorite. Several observations must be co-ordinated to avoid error as the apparent direction varies with the position of the observer.

Meteoriticist Scientist who is a specialist in the study of meteors and meteorites.

Meteoritics The science of meteorites and associated phenomena.

Meteoroid A small body of solid material in outer space. The average speed is 120,000 km/h, but cosmic velocity can be as high as 150,000 km/h. The speed upon entering the atmosphere depends on whether the meteorite is overtaking the Earth in the direction of its rotation or not. Ninety-nine per cent of the atmosphere is less than 40 km deep.

A piece of cosmic material of cometary or astroidal origin in orbit around the sun. A meteoroid that has entered our atmosphere and generates light, either from the glowing of the body itself, caused by friction, or from the luminescence of the glowing gases surrounding the body. Technically, this light is the 'meteor'.

Meteor Shower is a group of meteors, all burning at the same time, produced by the break-up of a large meteoroid. These showers are sometimes called 'swarms'. A shower may be composed of anywhere from less than ten to many hundreds of fragments. Individual small meteorites in a shower are generally distributed over an elliptical area.

Meteors 'Shooting stars' with visible luminous trails due to the friction with air when a meteorite passes through the atmosphere. The body itself is called a meteoroid and, when recovered on earth, we call it a meteorite. A fireball is a large meteor.

Milky Way Now used by astronomers as the name of our own galaxy. Shaped like a disk, the galaxy appears as a band across the heavens.

N

Nebula Any blurred or diffuse light source in the sky. Although the term was once applied to objects both inside and outside the Milky Way, it is now commonly restricted to gaseous dusty clouds inside the Milky Way — for example, the Crab nebula.

Neutron Elementary particle that is one of the building blocks of the atom's nucleus.

Neutron Star Small but massive collapsed star composed entirely of neutrons.

O

Octahedrite Most common type of iron meteorite, having an octahedral structure. Contains bands of taenite and kamacite referred to as Widmanstätten structure.

Octahedron A solid geometric form having eight faces.

Open Universe The view of some cosmologists that the universe is destined to expand eternally.

Orbit The path of a body that is in revolution about another body, as is the Earth about the Sun.

P

Pallasite Stony-iron meteorite in which the stony portion consists of olivine.

Path The projection of the trajectory of a meteor against the sky as seen by an observer.

R

Radio Astronomy The branch of astronomy that deals with the study of radio sources, which can be planets, stars, galaxies or quasars.

Regmaglypts Depressions resembling thumbprints, found on the surface of many meteorites, produced by ablation.

S

Siderite A meteorite composed mainly of iron-nickel alloys.

Smoke Trail Observable trail of meteorites falling during the daytime. A fine dispersion of meteoritic matter can be seen in the sky after the meteorite has fallen. In the case of a large meteorite shower, the trail can even obscure the Sun for some time.

Solar Wind The steady stream of charged particles coming from the Sun.

Space Travel and Meteoroids It would take a journey of several hundred Earth years, travelling at almost the speed of light, to stand a chance of finding a technologically advanced community in space.

No sources of energy have yet been developed, or even conceived, that could provide propulsion for attaining anything like 99 per cent of the speed of light. Another factor to be reckoned with is any impact with a meteoroid at these speeds. It has been estimated that a small grain of meteoroid material colliding with a spacecraft at near luminal speed would have the effect of an atomic explosion. Scientists whose work borders on science fiction dream up solutions to these problems, such as a laser that could vapourize meteoroids before impact.

Sunspot A darker, relatively cool patch on the Sun's surface.

Supernova A temporary star which flares and fades, occurring in our galaxy perhaps five times in a thousand years. As the centre of the star collapses, with temperatures a thousand times higher than that of the hydrogen bomb, the star's luminosity is increased a million times. It may shine as brightly as 200-million suns. The material is blown violently into space, forming a glowing nebula of very hot gas which can be seen without a telescope. In the past 1,000 years, only four have been recorded (1006, 1054, 1572 and 1604). The Allende meteorite could be supernova debris.

T

Tektites are not true meteorites. They have spent very little time outside the Earth's atmosphere, whereas meteorites have spent millions of years in space. The basic chemical make-up is similar to that of natural glass. It is generally thought that they were made by the impact of comets on the Earth, which caused rocks to evaporate and eject into space. These later condensed into droplets and re-entered the atmosphere.

The theory that they originated on the moon is almost completely discredited. They are considered as close relatives of meteorites, as they bear the unmistakable marks of a flight through the atmosphere.

Train Anything remaining along the

trajectory of a meteor or fireball after the head of the meteor has passed. May be light, dust, vapour, or ionization.

Trajectory True line of flight of a meteor or fireball relative to the Earth.

U

Universe All of space, matter, energy and time.

W

White Hole The reverse of a black hole; instead of trapping energy and matter, it spills it out.

Widmanstätten Figures Figures appearing on an etched surface of an octahedrite. The result of an intergrowth of kamacite and taenite produced as the meteorite parent body cooled in space. Named after Alois von Widmanstätten, who discovered them.

When the cut and polished face of an iron-nickel meteorite is etched with acid, a very interesting pattern of crystallisation appears. This is the most striking feature of some iron-nickel meteorites.

Meteorites having this structure are called octahedrites.

X

X-Rays A form of radiation that lies between the ultraviolet and gamma-ray wavelengths.

Important Meteorite Collections

The American Museum of Natural History
Central Park West at 79th Street
New York, New York 10024

Center for Meteorite Studies
Arizona State University
Tempe, Arizona 85281

Institute of Meteoritics
The University of New Mexico
Albuquerque, New Mexico 87131

Field Museum of Natural History
Roosevelt Road and Lake Shore Drive
Chicago, Illinois 60606

Smithsonian Institution
14th and Constitution Ave, N.W.
Washington, D.C. 20560

McDonnell Center for Space Sciences
Washington University
St. Louis, Mo 63130

National Science Museum
Ueno-Park, Taito-ku, 110, Japan

Geological Survey of Canada
Canadian National Meteorite Collection
Ottawa, Ontario, Canada

Meteor Centre
National Research Council
Ottawa, Ontario, Canada

Earth Physics Branch
Department of Energy, Mines & Resources
Ottawa, Ontario, Canada

Meteorite Observation and Recovery
Project, Physics Department
University of Saskatchewan
Saskatoon, Sask., Canada

Naturhistorisches Museum
Wien, Austria

British Museum of Natural History
London, England

Muséum National d'Histoire Naturelle
Paris, France

Academy of Sciences
Meteorite Committee, Moscow

Max Plank Institut für Kernphysik
Heidelberg, West Germany

Mineralogical Museum of Copenhagen
University, Denmark

Museo Nationale
Rio de Janeiro, Brasil

Academy of Sciences
Warsaw, Poland

Australian Museum
Sydney, Australia

Bologna University
Italy

Ege University
Izmir, Turkey

Geologinen Komites
Helsinki, Finland

Geological Survey of India Museum
Calcutta, India

Hungarian Natural History Museum
Budapest, Hungary

Humboldt Universität zu Berlin
Germany

Further Reading

Calder, Nigel. *Violent Universe.* New York, Viking Press, 1969

Merrill, G. P. 'The Story of Meteorites', *Minerals from Earth and Sky,* Vol. 3, Part I, Smithsonian Scientific Series, 1929, pp. 1-163.

Farrington, O. C. 'A Catalogue of the Meteorites of North America to January 1, 1909', *Memoirs, National Academy of Sciences,* Vol. 13 (1915).

Farrington, O. C. *Meteorites* [published by the author], Chicago, 1915.

Hey, M. H. and **Prior,** G. T. *Catalogue of Meteorites,* William Clowes & Sons, London, 1953.

McCall, G. J. H. *Meteorites and Their Origins.* New York, John Wiley and Sons, Inc., 1973.

Dietz, Robert S. *Astroblemes,* Scientific American, Vol. 205, No 2. (1961), pp. 50-59.

Vagn, F. *Buchwald,* Handbook of Iron Meteorites. Univ. of California Press, London.

Buddhue, J. D. *Meteoritic Dust,* The University of New Mexico Press, Albuquerque, 1950.

Anders, Edward, *Origin, Age and Composition of Meteorites,* Space Science Reviews *3,* pp. 583-714, 1964.

Hawkins, Gerald S., *Meteors, Comets, and Meteorites,* McGraw-Hill, 1964.

Heide, Fritz. *Meteorites,* The University of Chicago Press, 1957.

Krinov, E. L. *Principles of Meteorites,* Pergamon Press, 1960.

Krinov, E. L. *Giant Meteorites,* Pergamon Press, 1967.

La Paz, Lincoln, *Space Nomads: Meteorites in Sky, Field and Laboratory.* New York, 1961.

Mason, Brian. *Meteorites,* John Wiley and Sons, 1962.

Middlehurst and **Kuiper,** G. *The Moon, Meteorites, and Comets.* University of Chicago Press, 1963.

Nininger, H. H. *Arizona's Meteorite Crater.* World Press, 1957.

Nininger, H. H. *Out of the Sky.* Dover, 1959.

Watson. *Between the Planets.* Harvard University Press, 1956.

Spencer, L. J 'Meteorite Craters as Topographical Features on the Earth's Surface', *Geographical Journal,* Vol. 81 (1933), pp. 227-248.

Perry, S. H. *The Metallography of Meteoric Iron.* U.S. National Museum Bulletin No 184 (1944).

O'Keefe, John A. *Tektites.* The University of Chicago Press, Chicago.

Boyd, George A. (Editor) 'The Published Papers of H. H. Nininger', *Biology and Meteoritics.* 1971.

Moore, Carleton B. and **Sipiera,** Paul P. *Identification of Meteorites,* 1975.

Lewis, Charles F. and **Moore,** Carleton B. 'Catalog of Meteorites in the Collections of Arizona State University', 1976.

Nininger, H. H. *Meteorites — A Photographic Study of Surface Features,* 1977.

Wood, J. A. *Meteorites and the Origin of Planets.* McGraw-Hill, 1968.

14.95